Digital Mythologies

Digital Mythologies

The Hidden Complexities of the Internet

----- Thomas S. Valovic

Rutgers University Press
New Brunswick, New Jersey, and London

Library of Congress Cataloging-in-Publication Data

Valovic, Thomas S.
    Digital mythologies : the hidden complexities of the Internet / Thomas S. Valovic.
        p.   cm.
    Includes bibliographical references.
    ISBN 0-8135-2754-6 (cloth : alk. paper)
    1. Information technology—social aspects.   2. Internet (Computer network)—social aspects.   3. Computers and civilization.   I. Title.
HM851.V35   2000
303.48'33—dc21   99-23161
                                                                                          CIP

British Cataloging-in-Publication data for this book
is available from the British Library.

Manufactured in the United States of America

To my wife, Elaine, and daughter, Jennifer

Contents

Preface

Since I consider myself to be a bit of a technophile, writing this book was an interesting experience. It certainly required me to think outside of the box. The genesis of the book was in large measure my need to reconcile what I perceived as a kind of logical disconnect at work in the stories about and descriptions of the Internet in the popular media. In general, the media seemed to have gotten caught up in the Internet craze, almost in a pop-culture sense, and they became prone to endlessly repeating a single idea: that the new technologies were going to profoundly change our lives in the realms of business, education, health care, and just about any other realm of human activity we could think of.

While this idea was true enough on the surface, a little analysis revealed that these breathlessly delivered renditions were usually devoid of any real specifics. In a sense, the media seemed to be involuntarily parroting the newly minted conventional wisdom without any deeper analysis or questioning of assumptions. I wanted to find the answers to some basic questions: What are the specifics of these transformations? How do we get past all the marketing hype and the constantly invoked mythologies of digital progress and into something more tangible, more grounded in day-to-day reality? If we suspect that something historically great or interesting or unusual is occurring, then why should we stop short of making a case for the human use and application of new technologies in terms of benefits derived? After all, if we do not know what these changes will really amount to, how can we possibly say that they are important or exciting or wonderful?

When I began to work on the book, there was definitely a void in the research being done on the social implications of the new digital technologies. Traditional academia was taking on some of these issues but often seemed more interested in focusing on the technology itself rather than in using their considerable resources in the social sciences and humanities to make projections about technology's effect on the quality of life. I found myself disappointed when I attended Harvard University's Conference on the Internet and Society in May 1996. Technology and business clearly were the loci of attention. On the one hand, the mantra of technology-driven business—we need to stay competitive—seemed to have worked its way into the hallowed halls of the academy as a part of what I call the "new utilitarianism." On the other hand, universities often had their hands full playing catch-up with respect to the latest technical advances in computers and communications. The corporate sector was at the white-hot center of technological growth, and the universities had to follow the corporate trajectory of development to keep current.

One explanation for the void in research is simply our inability to keep pace with the rapidly moving technology, let alone analyze its long-term effects; but the best explanation seems to lie in the fact that traditional social commentators were not up to speed on the technology and technologists were not oriented toward social issues. So into the breach went a group of technologists, representatives of digital culture claiming they knew how the Internet was going to change the landscape of contemporary life. *Wired* magazine, for example, told readers in its premiere issue that contributors were going to write about the social impacts of the technology. However, they ended up writing about its impact on the magazine's own narrow constituency, the digital elite. All that this group of admittedly articulate and technologically savvy writers and analysts really accomplished was to crank up the hype about how the new technologies were going to transform the future of society. Their doing so only exacerbated the original problem. The proponents of digital culture were long on

enthusiasm and on vague, effusive Emersonian assertions that were dis-
tressingly short on depth and specifics, and this encouraged the less rigorous representatives of the mainstream media to move into similarly
nebulous realms.

When I started this project, I did not intend for it to be quite so
focused on Net politics and technocracy, which had been addressed so mas-
terfully in Neil Postman's book *Technopoly*. However, I was working on
this project at about the same time that the defining events of the 1996
presidential election were taking place. Ross Perot emerged as the nation's
first full-fledged technocrat-as-candidate with his elevation of the computer
to the role of policy advisor and his curious talk about solving the nation's
political problems by using "pilot programs" and "debugging the system."
Many of the events of the election gave the incipient notion of electronic
democracy a new sense of urgency, and thus the movement toward technocratic governance took on for me a renewed and special relevance.

A linkage that I hope is established and reinforced in these essays
is that between the advent of digital media and changing attitudes toward
reading, literature, and the classical traditions of scholarship. I must agree
with social critic Camille Paglia that our educational horizons have nar-
rowed seriously and unacceptably and, absent a major correction, will
continue to do so. There is enormous value to be derived from the online
world, but the notion that the Net will replace what Robert Maynard
Hutchins called the "Great Conversation"—the ongoing narrative of scholarship and the great traditions of learning—is specious. Accordingly, I
have argued that the Net and other digital technologies should not replace
wholesale the methods of information dissemination now in place—and
this replacement is the fervent wish of digital culture—but rather should
augment and amplify our existing modalities. I am convinced that failure to encourage the latter would be a serious mistake, although even as
I write we are already heading merrily down the former path.

Digital culture plays a central role in this book. I hope the reader will
appreciate that the term "digital culture" is a necessary generalization and
will mean different things to different people. Most specifically, the term
relates to the kind of thinking found in *Wired* and *Mondo 2000* and to the

thought leaders—of various generations, I might add—who foster it. Certainly *Wired* and its coterie of bright and capable writers and editors play a large part in defining digital culture. Readers who consider themselves to be a part of digital culture may not see their own specifics mirrored in my observations. I hope my friends and colleagues who do consider themselves part of this culture will appreciate the necessity for my broad use of the term and my elastic interpretation of it.

There are many talented and well-intentioned people in digital culture, and I have gotten to know many of them through the WELL. I have nevertheless been rather critical of digital culture in this book, and my use of the term is meant to connote the embracing of a value system that I feel is fraught with problems. I have not pulled any conceptual punches in stating my conclusions, especially in the section entitled "Digital Culture." Although I consider Kevin Kelly, Stewart Brand, and Howard Rheingold to be intellectual adversaries, none of my comments is intended to reflect upon them personally. I have considerable respect for what Stewart Brand in particular has accomplished, the various organizations he has built, and the people he has attracted—even though, in my opinion, he has not remained true to his own stated vision.

I am aware that both the content and the strategic direction of magazines and online services change with the passage of time. But the reader should understand that my comments concerning *Wired* and the WELL, which are the basis for a number of essays in this book, are based on my own encounters with these information sources over a period of several years—a reasonable sampling period by any measure.

As we head into the new millennium, the Internet offers us some exciting prospects. However, it seems premature to bring out the champagne bottles and celebrate its success as one of humankind's greatest technological achievements. I prefer to err on the side of caution and approach the Net not with the self-congratulatory effusion of the runner at the end of the race but with hesitation and circumspection—with a sense of wondering whether we are equal to the task of shaping its destiny toward the goals of humanity and away from the seemingly ineluctable momentum of what Langdon Winner calls "autonomous technology."

There is no question that the virtual world has its attractions. But it also has limitations, and our experiment with it is an awfully large one to undertake with a riverboat gambler's stance toward the future of society and the world our children will inherit. I am convinced that we cannot afford the luxury that we have had with other technologies of saying implement first and ask questions later. Recognizing the imperatives of exponential social change and the increasing urgency of our environmental predicament, I suggest that a deeper awareness should inform all of our future technological endeavors, given their increasing power over social and economic conditions.

My biggest concern about the stance of the thought leaders of digital culture toward the Net is their lack of gravitas about the things they are trying to replace—the grand traditions of scholarship and learning that, as if tainted with guilt by association, are held to be lacking simply because they are represented in the medium that we call the book. This strikes me as a kind of superstition-in-reverse. It is all a bit too facile and trendy, and I do not quite know how to account for the stampede to the virtual other than viewing it as a strange media virus affecting otherwise intelligent and thoughtful people. The traditions of our civilizational narrative, even if they are flawed by deconstructed specifics of circumstance, should not be summarily abandoned for the sake of instantaneous access, lest we become, in the words of Paglia, "prisoners of contemporaneity." This kind of thinking is a bit odd, and a sense of balance is sorely lacking. While I truly believe that Internet technology has great potential to deliver many social benefits in the transformations that lie ahead, I wonder whether we might end up with an Internet—and a society—very different from the one that has been promised.

- - - - -

It takes a village to create a book. This book morphed through many incarnations, and I am immensely grateful to the individuals who played an important role in bringing it into its final—and in my mind most satisfactory—form. There were many who gave generously of their time or assisted in seemingly small ways that proved, in the final analysis, to assume much greater significance. I would like to thank CoraLee Whitcomb, Ken Goffman, Stuart Kaufman, William Irwin Thompson, Paulina Borsook, Kathleen O'Neill, Arthur Kroker, Thaisa Frank, Vinton Cerf, Gary Chapman, J. Baldwin, Mark Stahlman, Gail Williams, John Seely Brown, Avital Ronell, Douglas Rushkoff, Jack Trainor, Charles Firestone, and Hank Roberts.

I would also like to offer a special thank you to Jim Fallows, formerly editor of *U.S. News and World Report*. I owe a particular debt of gratitude to Jim for his support for an article that eventually formed the basis for the section entitled "Digital Culture." He provided an important validation for a thesis that had taken me into uncharted terrain. Also in the special mention category is the invaluable help afforded by Linda Garcia, formerly of the Congressional Office of Technology Assessment, who reviewed the book and provided her critical feedback.

In addition, I would like to thank the individuals on the WELL with whom I have over the years had numerous discussions, many of which were related to the themes explored in this book. More specifically, I would like to thank a group of individuals who provided interesting and valuable perspectives during the research phase of an article I wrote for *Media Studies Journal* on the dynamics of online media criticism. Many of these individuals I came to know through the WELL's media conference, and much of the material from these discussions eventually found its way into essays in the section entitled "The Electronic Polity." The group includes John Seabrook, Peter Sussman, Howard Rheingold, Paula Span, Philip Elmer-Dewitt, Kevin Hogan, Roger Karraker, Tom Mandel, Art Kleiner, David Gans, and Jon Carroll. I would also like to thank Professor Jerry Berger of Northeastern University and Rob Snyder and Everette Dennis of the Media Studies Center for encouraging my initial explorations of these areas.

In terms of the honing of the manuscript itself, my former agent, Richard Curtis, provided necessary and useful structural advice on early versions of the book. I would also like to thank literary counsel Nina Graybill for helping me navigate the complexities and snags that sometimes emerge on the long road to publication. Most importantly, I want to extend my thanks to Helen Hsu, my editor at Rutgers University Press, for her positive and professional attitude and for managing to find that wonderful middle ground between encouragement and constructive criticism. Her comments and those of the reviewer for the book were extremely helpful in giving it a conceptually sound framework. Jack Repcheck of Princeton University Press played a strong enabling role in getting the manuscript into the right hands. Carlotta Shearson did a first-rate job in preparing the manuscript for publication.

Finally, I would like to thank my wife, Elaine, and daughter, Jennifer, who cheerfully put up with the many lifestyle changes that writing a book entails. Elaine helped in innumerable ways, large and small. Her tenacity, helpful insights, and skills as a sociologist, researcher, and first-rate Web surfer were indispensable throughout the course of the writing and research for this book.

Tom Valovic

Holliston, Massachusetts

December 15, 1998

Digital Mythologies

Virtual Dreams

Virtual Folklore:

Breaking the News

about the Internet

Inter-what? Never heard of it.

Isn't that the fancy government network that's been around for years? Academics use it. . . . So what's the big deal?

Let me know when it's real; sounds like vaporware to me.

Nobody seemed to understand what all the fuss was about. Being the first to break a story can have its disadvantages. First, there is no one to validate your discovery; so you know that professionally you are about as far out on a limb as you are ever going to get. You shift around a lot psychologically and never quite get comfortable with your cherished piece of information. Second, you find yourself slipping into missionary mode a bit too often, occasionally exhibiting the strained, too earnest quality of a UFO abductee: "Yes, I know that no one else heard the tree falling in the forest. But I did, I was there, it was *real.*" You know you have to avoid making the story sound like too big a deal, sounding too earnest and, well, scaring people away.

But in my case, the falling tree was a giant redwood: the beginnings of a commercialized Internet, not only a new technology but a major paradigm shift in the technological landscape. My magazine, *Telecommunications*, had gotten the scoop, and I was the editor who had snagged the story. With the pain and difficulty that only a steep learning curve can inflict, I found out that it is indeed possible to get a scoop too early. Spreading the news was a bit like lecturing in a large hall when only twenty or thirty people have shown up. The problem was that there was no critical mass and no way to produce it. Even if I had appeared on CNN and trumpeted my findings to the global village, no one would really have known what I was talking about. (Well, Vinton Cerf, considered by many to be the father of the Internet, would have known, of course; and so would the small number of academics who understood—really understood—what was about to happen.) No, this kind of news would take time to develop and to be fully appreciated.

I broke the story about the Internet's move into the commercial market in 1991, a full two years or so before the mainstream press locked onto the Internet like a heat-seeking missile in 1993. As senior editor of *Telecommunications*, I was in possession of the biggest communications story of the decade, trying to figure out what to do with it.

The saga, fraught with many twists and turns and its own particular brand of journalistic angst, began one perfectly ordinary Sunday afternoon in 1991 when I received an e-mail message that would rock the foundations of my professional and personal life for the next year or so. The message was from Brian Kahin, who headed up Harvard University's Information Infrastructure Project. (Kahin would go on to play a key role in the Clinton administration's Office of Science and Technology Policy.) Kahin had sent me a classic news tip: The Internet, which had up to that point been largely a behind-the-scenes academic network, was about to be let loose as a commercialized entity. Why Kahin had chosen to leak this news to me I was not sure, but I knew that I had a journalistic tiger by the tail as I gazed at the message scrolling across my laptop's silver-blue screen.

Back in 1991, the Internet was the world's best kept secret. Its primary users were academics and defense contractors, and it existed in a world apart from the telecommunications industry. For the most part, even the more knowledgeable and savvy telecommunications experts did not pay much attention to it—if they knew that it existed at all. The same, amazingly enough, applied to most of the senior executives of the Regional Bell Operating Companies, or RBOCs, as they came to be known.

Kahin's e-mail was a blockbuster because it addressed the incipient commercialization of the Internet. This meant that the Net was about to be unleashed from the relatively narrow confines of academia and a hitherto slumbering beast was about to wake up and conquer the world.

Internet commercialization would mean that private industry could develop and market Internet technology as aggressively and effectively as it had personal computers. This event would eventually transform the Internet into a publicly available network that could be used by anyone with access to a personal computer and a telephone line. This breakthrough was going to massively transform communications and most other areas of human endeavor as well, including business, education, science, and entertainment.

Kahin's e-mail put all of my cognitive resources on red alert. Could it be that I had the scoop of the decade in front of me? I had to find out, and I began researching the story in the most logical and convenient fashion available—by using my networked computer. The first task was verification. I initiated an extended e-mail conversation with Kahin on the significance of the commercialization angle. Eventually, he convinced me that this was no mere speculation and that commercialization was imminent. But I needed more perspective, and I also needed to make sure I did not inadvertently tip my hand to competing publications in the course of doing the background research.

To continue reality-testing the story, I had some excellent sources available. One of them was Vinton Cerf. Cerf was a member of *Telecommunications*'s editorial advisory board. He had also written the foreword to my book *Corporate Networks* and had recommended it for publication. During a recent visit to the magazine's parent company, Horizon House, Cerf and I, along with publisher Jim Budwey, had delved into the complexities of Internet commercialization over plates of Duck Choo Chee and Pad Thai. Cerf, nattily attired in his usual dark three-piece suit, had confirmed for me the scope and importance of what was happening.

Even after I had Cerf's confirmation, my work was not finished. I had to make sure that the story was airtight, impeccable, completely solid. I gathered my thoughts, based on my conversations with Kahin and several other experts in the field, and e-mailed a version of the story to Cerf to get his reaction. Interestingly, I realized that throughout the entire process I had inadvertently conducted my first experiments with Net-enabled journalism, which felt qualitatively different than the usual means of generating a story. The process felt more academic, more collaborative—a bit like peer review. Whether the approach was qualitatively better than traditional journalistic methods became an interesting question, one that is still being discussed by computer journalists and others who have engaged in similar experiments.

In addition to the draft story of the commercialization itself, I was also working on a fascinating angle that I was convinced would be almost as important as the main story: The commercialization of the Internet appeared to be happening without the awareness of Washington's telecommunications policymakers, including key officials at the Federal Communications Commission (FCC), whose job it was to monitor such developments. Not only was this development not on official Washington's radar screen, but most of the telecommunications industry itself was blissfully ignorant of its snowballing importance.

After about a month of intense online and offline research, the story was finally coming together. As production deadlines neared, I was faced with the critical and difficult decision of whether to publish it. There were several cautions and caveats to consider, and the story would have to be framed carefully. First, from a technological standpoint, the Internet was fairly new territory and not something that our magazine (or any of our competitors, for that matter) had generally covered. As an academic rather than a business network, the Internet's relevance to our readers seemed quite limited. As a result of this limited coverage, we did not have the luxury of a large experience base to draw upon in launching the story, and there were few traditional ways of verifying it. Technologically and journalistically speaking, I would be out on a limb. After due consideration, however, I decided that the research was solid enough to assume whatever risks there were, and we went with the story.

The item first appeared in an article entitled "A Giant Step towards Internet Commercialization?" in the June 1991 issue of *Telecommunications:* "According to several sources, ANS [the backbone operator of the Internet] is expected to make a major announcement concerning the creation of a separate for-profit subsidiary chartered to develop the T3-based Internet backbone for commercial use. If this does happen, it will have

significant impacts for further Internet commercialization, according to
Brian Kahin, Director of Harvard University's Information Infrastructure
Project."[1] The piece went on to point out that, as we have seen, Internet
commercialization appeared to be proceeding without the sanction or
awareness of any of the federal watchdog agencies that normally oversaw
major developments affecting national telecommunications policy. I had
managed to confirm this fact by, among other things, interviewing the
FCC's Robert Pepper, one of the agency's principal policy advisors.

There were several ways to look at this apparent lapse. One reaction
was to say "So what?" The Net was such a new development, and most
people could not be expected to ferret out its the full implications. But there
was another possible perspective: to see what was happening as a kind of
technological coup by a small number of elite experts who did fully fore-
see the implications of unleashing this particular genie.

Let me be careful in my phrasing here. The plan to commercialize
the Internet clearly was not a conspiracy to overthrow the nation's telephone system, which was bureaucratized, unfriendly to data traffic, and
notoriously unamenable to change; but the move did have several important and fairly predictable outcomes, one of which was competition for and
even displacement of many functions of the nation's existing communications infrastructure. While displacement seemed far-fetched at the time,
even among the most vociferous Internet advocates, it has indeed happened
in selected areas. Furthermore, a displacement of the phone system as a
whole is now being seriously discussed with the advent of Voice-over-IP
(VOIP) technology. There are those who think, in other words, that the public telephone network has become a dinosaur because of the Internet.

In 1991 there was mounting frustration within the computer community over the fact that the telecommunications industry was so heavily regulated, and this frustration was accompanied by the conviction that
this regulation was hampering technological development in a variety of
areas. Thus, the commercialization of the Internet was, and still is, construed by some as an end run around the policy establishment in Washington and the traditional hegemony of the public network. In a purely
technological sense, it was a preemptive first strike.

I found these developments quite stunning. How could a twenty-megaton event like the arrival of the Internet occur without being noticed by the nation's most plugged-in and savvy experts in the telecommunications industry? How could the FCC be looking the other way when one of the most important telecommunications capabilities invented in the twentieth century made its debut?

A good starting point for understanding the stealthiness of the Net's initial introduction is an appreciation of the huge cultural and content chasm that has traditionally existed between the computer and telecommunications industries, two distinct technical communities. Even though the Internet would eventually be perceived as a vehicle for telecommunications, the technical development of the Internet was accomplished largely by the computer community. The basic technology that made the Internet possible is a data communications interoperability protocol called the Transmission Control Protocol/Internet Protocol, or TCP/IP. This standard, developed by Vinton Cerf and Robert Kahn, allowed disparate types of computers to communicate with each other over the telephone lines. As a "technological coup" of rather extraordinary proportions, Internet commercialization took place without the knowledge of major executives and managers within the telecommunications industry. With the possible exception of Bell Atlantic, the regional Bells were largely unaware of not only the fact of commercialization but, more importantly, its practical business implications. Few telephone company executives anticipated how private companies could be given the charter to expand the Internet on both a national and global basis in such a way as to compete with the RBOCs's huge and well-established networks. By the time commercialization was more or less a fait accompli, the regional Bells and the rest of the telecommunications industry began to realize its significance. But by then they were playing a serious game of catch-up.

Once the Internet had gathered enough momentum, its growth was out of any one entity's hands, and it became, quite simply, unstoppable—all in all, a remarkable progression of events. Once the Net was commercialized, it simply became a matter of users and companies signing on with a service provider—and new ones sprang up every day. In retrospect, it is possible to speculate that if the RBOCs had better understood the Internet phenomenon and realized its potential to become a universal network for all forms of communication—possibly also including voice— they might have tried to squelch or gain control of the process.

About three months after my article appeared in *Telecommunications,* the *New York Times* came out with a similar story describing the process of Internet commercialization. When I saw the item, written by technology writer John Markoff, I was taken aback. Several months before the item appeared, I had faxed a copy of the *Telecommunications* piece to Markoff. To this day, I have not been able to establish whether the *Times* used my article as the basis for their story.

Six months or so after *Telecommunications* ran a series of articles on Internet commercialization and its potential effects on the industry, various other trade publications, such as *Communications Week,* began covering the story. Our monthly had, in essence, scooped the weeklies by about six months.

In the years that followed, I tracked the Net's elaborate progressions both as a technological phenomenon and as a transformative cultural force of unique proportions and capabilities. As editor in chief of *Telecommunications,* I had a special vantage point from which to view these developments and found irresistible the urge to record the picture in the manner in which I saw it. The essays in this book are largely born of my notes and observations of the Net on its special trajectory through the core of our cultural imagination.

Point of No Return:

Crossing the

Virtual Threshold

There is an unearthly glow coming from the mock-cavernous and hopelessly disorganized living space that is my den. The pale glow is emanating from the small screen, approximately seven inches by ten inches, of my laptop computer. The glow beckons, like a window to another world. It seems inconsistent with everything else in the room, as if it were somehow oddly apart from the world of physical things that we inhabit. As I glance at it, it vaguely reminds me of the rush of spirit that was channeled through a television set in Steven Spielberg's movie *Poltergeist;* indeed, that pale glow almost seems to partake more of the spirit world, or at least some less immediate, less tangible dimension than its more mundane surroundings. The glowing window beckons seductively. The odd thing is that the bluish-white glowing screen is indeed a window to another world.

What is this other world? Am I being coyly metaphorical or ploddingly literal? This other world is, of course, the virtual world, a realm of almost Platonic purity that is not matter and not mind but disturbingly in-between. In form and in likeness, the virtual world resembles the physical realm that we normally inhabit; but it is also different in ways that challenge our normally invoked powers of description. In fact, in order to zero in on the more elusive philosophical implications of this new dimension of existence called cyberspace, we often find ourselves forced to resort to poetic usage to describe its almost numinous qualities. At the center of this new virtual world lies the swirling complexity of the Internet—a huge, amoeba-like, and chaotic thought cloud of human interaction that constantly grows and changes like some out-of-control lab experiment.

As the Internet evolves, we laggard and worn-out humans are no match for the swift and unmerciful efficiency of silicon. We seem somehow compelled to try to evolve, to change ourselves—as well as our modalities of perception and our sense of societal direction—in order to keep up. As a Hollywood producer might say in whispered tones to an associate: "This is big."

But just jumping on the bandwagon is entirely different from knowing where the parade is headed and how it will affect the more mundane intricacies of everyday life. After a while, we begin to suspect that some serious analysis is in order. The Internet is no longer a new phenomenon. And yet our analysis of it as a unique social and cultural force of transformation seems to lag behind its trailblazing trajectory into the next century.

Why should this be the case? The answer to this question lies partially in my playful schema for describing the response to any given new technology. In this schema there are three phases: information euphoria, information ennui, and information wisdom. When it comes to the Internet, we are still in "information euphoria": This curiously contagious sense of enthusiasm for all things digital is still rampant, and, while seemingly harmless, it has been known to impair critical thinking in the sharpest of minds. The current hegemony of the Net in everyday life and its rapid adoption in the space of only a few years are indeed impressive. It is easy to be awed by the seemingly spontaneous generation of a worldwide system of communications, planned and implemented by no single corporate or governmental entity and yet, like a force majeure, swiftly and autonomously extending its reach into all levels of life and all corners of the globe.

However, analysis of the deeper cultural transformations being wrought by this technology has so far only scratched the surface. The process of scientific and humanistic discovery by which the full implications of the Net's effect on everyday life is a long and winding road. But eventually information euphoria, having run its course, gives way to a more realistic perspective: that there are deeper issues underlying the Internet phenomenon, issues that, in some cases, even expert commentators and observers have overlooked.

Cyberspace is experiential. As such, it yields different results for different users. This property creates unique challenges in terms of empiri-

cal observation and generalization. In one class, we have the true believ-
ers, who enjoy the numinous quality of cyberspace and are becoming
increasingly comfortable in virtual spaces. Some may even invest more
of their energy and awareness in these artificial worlds of pure mind play
than in the physical world of Wal-Marts, Seven-Elevens, budget cutbacks,
gangsta rappers, militias, and all the other ragtag iconography of postmodern life. Perhaps, in some way not yet wholly understood, these individuals have found their own "private Idaho," a cyber-Valhalla, or an
entry into Wallace Stevens's "worlds of fictive imagination"—pristine electronic spaces free of the unpleasant side effects of civilizational entropy.
In this sense, cyberspace, like television, can be addictive, offering an
enticing kind of sensory and cognitive stimulation. It is out of this kind
of experience that digital mythologies are born.

Unfortunately, this view of cyberspace, while quite prevalent,
tends to be one-dimensional. The media-dubbed doyenne of cyberspace,
Esther Dyson, may temporarily enthrall and tantalize us with her particular brand of information euphoria; or a Nicholas Negroponte, another
icon of the digital age, may convince us that we are indeed part of a
technological progression that historically surpasses all others. But eventually we will be left with the unsettling emptiness of questions left
unanswered or, worse, never asked at all. These purveyors of the most
sophisticated kind of digital culture never challenge basic assumptions
about the technology simply because no impulse to question them ever
seems to arise.

The reality is, of course, far more complex and obscure. The fact is
that cyberspace as an experiential phenomenon is riddled with hidden
complexities. It is, where discourse is concerned, a decontextualized, mul-
tidimensional environment in which the normal cues for detecting the sub-
tleties of human communication are often lacking. It is a virtual space
where reality itself might be part of an evolutionary design that employs,
among other things, the art of existential masquerade.

The impulse to simplify Internet issues is clearly a by-product of
the postmodern state of grace called information euphoria. The advent
of the Internet as a commercialized entity has been characterized by a

surfeit of hype that has badly clouded the public dialogue. The media hype can be traced to a number of sources.

Some of the hype stems from the strident renderings of the proponents of digital culture in their enthusiasm for promoting their new lifestyle centering on the virtual world. Some of it can be traced to the providers of information-age products and services, who have a vested interest in promoting the notion of some vaguely defined major societal transformation. In this sense, the sometimes superficial but always well-amplified posturing of *Wired* magazine concerning the techno-utopian future is a message that these providers are only too eager to borrow. Thus, there is a nicely symbiotic relationship between digital culture and corporate interests intent on defining the emerging landscape of the digital age. Yet another source of the hype is the media, which have tended to rely largely upon the oversupply of readily available information from the first two sources rather than upon the deeper digging necessary to unearth the full implications of the new technology.

As extensive coverage of news about the Internet began to increase, media commentators, some of them technically savvy and others less so, struggled to determine the societal significance of this new technology. What were its implications for business? For education? Would the love affair with the Internet amount to anything more than infatuation, a kind of episodic "geek love"? Would large numbers of Americans enthusiastically jump online and kindle a new electronic democracy, the beginnings of a bold new social experiment? Would the Internet further exacerbate the already sharp divisions between the information-rich and information-poor (not to mention the money-rich and the money-poor); or, on the contrary, would it create new synergies of wealth creation that would eventually usher in the long promised fulfillment of a techno-utopian society of leisure in which we are all watched over by "machines of loving grace," to quote poet Richard Brautigan?

By the time full-scale Internet-style information euphoria had set in, there was no hyperbole that could not be invoked with impunity. John Perry Barlow, cofounder of a lobbying organization called the Electronic Frontier Foundation (EFF), blithely informed us that the Net was the greatest invention since "the capture of fire." *Wired* developed hype-genera-tion into an art form with its unique way of casting the starkly technical in quasi-poetic terms. In fact, in several particularly grandiose articles that sought to fuse metaphysics with computer logic, *Wired* did not hesitate to render a vision of the Net's vast global interconnectedness in almost mys-tical terms. The vision worked as long as you did not ask too many ques-tions or otherwise probe too deeply.

Even though in the right context there are compelling justifica-tions for information euphoria, it is often invoked for the wrong reasons or on the basis of incorrect assumptions. For example, a common idea (or perhaps, more accurately, mild delusion) prevalent in digital culture is that all forms of important knowledge and information will eventually reside on the Internet. The notion is based on the projection that one day, with a few clicks of a mouse, a vast array of the world's information and knowl-edge—the virtual apple of all knowledge—will be ripe for the picking. Despite the fact that this notion has little basis in reality, it crops up con-stantly in the mythos of cyberculture, in the conversations of technocrats, and, just as often, in various computer industry venues. A classic example involved a statement Larry Ellison, CEO of Oracle Corporation and one of the more well known captains of Silicon Valley, made about the Inter-net during one of his speeches: "It's collecting all the knowledge of mankind and making it available in digital fashion—reliably, securely, and economically."

The second phase in my schema for expanding the horizons of our thinking about cyberspace is "information ennui." If information eupho-ria is the simulacrum of a drug-induced high, then information ennui is the inevitable hangover, the karmic payback. During this phase, we grad-ually become aware of what might be called the dark side of information technology. Some of the best scientific and technological inventions in his-tory have been subverted for less-than-noble purposes either by the inexor-ability of market entropy or by the fact that two major forces that we associate with democracy—quality of information and enlightened plu-ralism—do not always pull in the same direction. On the other side of "energy too cheap to meter" lies a Chernobyl.

Information ennui is not only necessary but healthy, since a certain skepticism concerning our best inventions promotes the kind of vigilance that can help prevent societies from doing bad things with good technology. It is important to develop a sense of how the application of technology really works in the nonlinear progression that we call history. Rather in the manner of the Hegelian dialectic, a balancing act of sorts is involved. When positives and negatives are taken into account, a conceptual tug of war invariably develops over the ideological justification for deploying powerful new technologies. These justifications are tempered in the heat of academic and political argument and counterargument, as well as to some extent in the judgments of the public.

Most importantly, information ennui is also characterized by a deeper questioning of the efficacy and social value inherent in emerging computer and communications technologies. Between the extremes lies the gray scale on which truth resides. In the case of the cyberspace debate, the thought leaders of information ennui are easily identifiable via a series of books that are highly critical of the digital revolution, including Clifford Stoll's *Silicon Snake Oil,* Sven Birkerts's *The Gutenberg Elegies,* Mark Slouka's *War of the Worlds,* and Stephen Talbott's *The Future Does Not Compute.*

Since these works do adopt extreme and highly polarized positions, their usefulness is at once accentuated and suspect. While useful in defining the borders of the digital debate, these positions suffer from the same one-dimensionality that cyberculture exhibits when it touts digital supremacy as a solution for society's problems. What is needed is reconciliation of these polarized positions on the basis of open and honest intellectual inquiry. This inquiry should reflect a few basic realities, such as the fact that most of us, with the possible exception of the Unabomber, depend on some form of computer and communications technology and would readily acknowledge that there are undeniable benefits involved in that use.

Like critic Sven Birkerts, I am puzzled by the fact that so few questions have been raised thus far about the implications of the emerging electronic polity, not only by cyberspace advocates, who have the capacity and perspective to do so, but also by our university-based humanists and social critics, who are nowhere to be found in this particular debate.

A minimally rigorous intellectual response to either position involves asking questions about both the primary and the secondary effects (not to mention the unintended consequences) of the rapid spread of new technologies, what I call the "technological diaspora." In a democratic society, where values must be constantly scrutinized, it is especially important to look at how these new regimes might alter our existing social and political structures, either for better or for worse. The results of this rigorous examination are, first, a finer and deeper appreciation for the complexity involved, the ability to differentiate between applications of technology that might be beneficial and those that might be harmful (a notion that is anathema to digital culture) and, second—perhaps most importantly— a fuller appreciation of the fact that the implementation of technology always involves trade-offs.

In my playful schema, the third phase of thinking about cyberspace eventually emerges from this dialectic. Progressing from the conditioned reflex that is information euphoria and on through information ennui, we arrive at our final destination, what might be called "information wisdom"—a pattern of thesis, antithesis, and synthesis. This phase has been tempered by a bit of brooding about the ways in which a fundamentally empowering technology either can be coopted for purposes of social control or can devolve by means of the process of commercial entropy into a pale shadow of its former self, just as some believe that television ultimately became what former FCC chairman Newt Minow called a "vast wasteland."

Information wisdom then is characterized by a healthy ambivalence toward new electronic technologies, particularly those described as powerfully transforming or far reaching in their social consequences. In this phase, an appreciation for the complexities of cyberspace emerges, as does an ability to see through the veneer of the oversimplified digital mythologies so ardently promoted by the thought leaders of digital culture.

Far too many treatments of the Internet deal with it as a phenomenon in a vacuum, wholly apart from the practical concerns of everyday life. The Internet, along with the digital revolution, is often viewed as an almost self-contained virtual universe with its own set of laws and modalities.

Wired magazine, for example, tends to treat the Net as an isolated phenomenon, almost as if the physical universe and all of its messy complications existed in a curious state of suspended animation. With an almost neo-Gnostic aloofness, *Wired* celebrates the disconnection of a technology from everyday life. Birkerts captures a sense of this virtual isolationism nicely in *The Gutenberg Elegies:* "The remarkable thing about *Wired* is that it presents a fully self-contained order, a closed circuit. Nowhere in its pages do we find any trace of the murky and not-so-streamlined world that we can still see outside our windows. No ice on mittens, no fumbling for quarters while the bus (late!) toils towards us through morning traffic. Everything is as clean as a California research park."[2]

Parallel universes are certainly enticing, as any science-fiction fan knows. But those who opt in to these universes are still opting out of society in some fashion yet to be fully defined. This has larger implications that need to be explored. And to speculate about what widespread usage of the Internet is going to mean for society, we clearly must jettison the tunnel vision of looking at the Net in a vacuum.

While the collective dialogue about the Internet is clearly still taking shape, that dialogue badly needs to be broadened in scope and recontextualized. One dimension of recontextualization means looking at the Net not as an isolated system but rather as part of a complex array of associated technologies that are currently under rapid adoption, as part of the technological diaspora. These technologies include videoconferencing, multimedia, satellite systems, wireless communications, voice messaging, and other items in the ever expanding mix of computer-based products and services finding their way into our daily lives. Another aspect of recontextualization involves looking at the cumulative effects of the technological diaspora on the fabric of daily life in a variety of contexts, including the personal, social, recreational, and professional spheres. Thinking about the Net as a separate system or phenomenon unto itself is relatively easy. When we examine the Net's many impacts on human cognition and habit, we can see it more holistically, as well as envision its long-term consequences. The final element of recontextualization then is completing this mosaic by once more widening the field of view to encompass a redefined role for science and technology in a postindustrial age. Completing the mosaic requires that we look at how the end products of technology might be used—either directly or indirectly, intentionally or unintentionally—to fulfill social objectives (the very essence of the notion of "technocracy") and at how these new complexities affect the process of information exchange, deliberation, and action that is the basis for a modern democracy.

The landscape of our cultural heritage is already changing in ways that will irreversibly alter the world that our children and grandchildren will inherit, although these changes have not yet come into focus. Insofar as it is linked to new systems of corporate and political influence, the role of science and technology in the late twentieth century is indeed a powerful one, and it should not be underestimated. Neither should the Internet's paradigm-shaping influence within this domain. As Alvin Toffler pointed out in *Powershift:* "The emergence of religio-politics around the world may seem to have little to do with the rise of the computer. But it does."[3]

Once the Net is recast in this larger cultural context, many questions arise. For example, if we embrace this new technology (as opposed to taking Birkerts's advice to simply "refuse it"), then what exactly are we embracing? Is this a package deal? If we assume that all technologies involve social and cultural trade-offs, then what might we be giving up in accepting the Net and the powerful capabilities of the electronic polity? Can we accept the interesting and exciting promise of the virtual world without the technological regimes and societal entanglements that accompany it? In other words, if *Wired* is not about technology but about lifestyle, are we free as a society to accept technology without lifestyle?

Examining the hidden complexities of the Net steers us toward still emerging dimensions of postmodern social structures and worldviews—shaped by the immersive sea of electronic media more than we perhaps care to admit or are consciously aware of. The message of these media is, in part, that the new technology is as much existential metaphor as practical tool with which to engineer new realities and immediate benefits. Certainly in one particular technology, virtual reality, we can see the expression of a powerful new metaphor that has the potential to change the way we look at ourselves and the world. Some observers, such as new-age commentator Terence McKenna, have even suggested that virtual reality has its roots in other, more traditional (and even ancient) types of purposeful and ritualized consciousness alteration.

If nothing else, the Net will surely continue to surprise us. The Net idealism that is at the heart of digital culture's somewhat naive enthusiasm may yet cause "a thousand flowers to bloom." But it will probably do so in ways that none of us anticipate, following the classic but unpredictable patterns of historical *enantiodromia*. Most significantly, we should not rule out the possibility that the stateless, borderless Internet will also provide models and metaphors for forms of social organization yet to emerge. It has clearly done so in the business realm, where human enterprise has been shaped by new models of collaboration and cooperation.

This brings us into the realm of Net politics and electronic democracy. Although I will address these subjects in the section entitled "The Electronic Polity," I would like to bring up the intriguing possibility that the Net might be a model not so much for new political structures as for a new cultural matrix, from which future political or even apolitical forms might emerge. Only faint glimmerings of this new structure can be discerned through the haze of postmodernism in the prospects for "hacking" consensus reality, an ontological inquiry that several decades ago began in a far different context. If the vigorous clash of ideas that constitutes the anarchic pluralism of the Net can ever congeal into a value system (and not just an online phenomenon), then such an event may provide some much needed clues toward solving the essential sociopolitical conundrums of our age: cultural relativism and epistemological gridlock. The Net may, in this sense, be the ultimate postmodern phenomenon.

I do not wish to discount the role of politics, however. If it is possible to imagine even more lateralization than the great American experiment in democracy has thus far afforded, then the Net could, in theory, provide a conceptual platform for such thinking to evolve. This, of course, makes the rather large assumption that the Net itself as a system of practices and a community can in the long term move beyond the luxury of chaos and into the stabilization of norms and mores. However, as most Net denizens will be happy to point out, this is a tall order. Digital culture cannot prize the anarchic and chaotic qualities of the Internet above all else and yet expect some kind of pluralistic cultural system or even model of governance to arise from those qualities. It cannot, in other words, have it both ways.

In the best-case scenario, the discourse of the Internet could foster a deeper understanding of cross-cultural values, which has so far been an elusive goal. With this understanding might come a nuanced appreciation of the more arbitrary qualities of the human condition. And, in addition, it might reinforce the notion that we individually create our own realities (while society mass produces them) and the notion that separate realities of equal value and validity can and do coexist. The Net may have the potential at least to become a vehicle for reconciling the postmodern Pandora's box of jarringly subjective truths in so far as they can be reconciled. Or, alternatively, perhaps such a notion is a pleasant chimera that leads us toward an impossible-to-manage cultural entropy.

As we slouch in Yeatsian fashion toward the end of one millennium and the beginning of another, our society seems to be suffering from a curious kind of option shock: so many choices, so many possible directions. Out of the postmodern mélange of raw potentiality that is the legacy of the 1990s, the Net summons us forth. It seems to promise to transport us into another dimension, place, state of mind, or realm of the possible.

As Sharona Ben-Tov has pointed out in her book *The Artificial Paradise,* our science fiction is in many ways our true cultural mythology. It is a particularly American phenomenon (one senses that the Soviets, unlike the Americans, were attracted to space less for its intrinsic fascination than for merely competitive purposes). In any event, the power and the glory of the Net is the obvious fulfillment of that mythology, at least on some level of collective awareness. What remains ahead is the reconciliation of that unique mythos with the deeper mythologies and structures of meaning that are embedded in the furthermost reaches of human awareness. What that reconciliation will consist of and how it will shape our final destiny in the coming decades remain to be seen and (perhaps) cherished as a new wave of human awareness.

When Cultures Collide:

The Internet versus

the "Great Conversation"

When I first heard Oracle's Larry Ellison utter the notion that the Internet would eventually embody all human knowledge, I could not decide whether I was hearing marketing language gone completely awry or a warning that we were headed for some strange times indeed as we move into the next millennium. As it turned out, Ellison was not the only one in the land of the virtual to be bantering this idea around. After poking around the nooks and crannies of digital culture and then listening to the urgency with which President Clinton and Vice-President Gore declared that all American schools should be wired to the Internet by the year 2000, I began to realize that there were plenty of responsible and thoughtful professionals and commentators who believed that there was a large basis in truth for such an assertion.

American educator Robert Hutchins refers to the human dialogue through time in history, arts, letters, and science as the "the Great Conversation." There are those who argue that the very essence of human learning is now undergoing a vast change in character. The gist of this argument is that the focus of human dialogue is now moving from traditional media—including books and the other appurtenances and tools of scholarly research—into the electronic realm epitomized by the Internet.

This, at least, is the contention of the proponents of cyberculture and the new technocrats, who often bolster their arguments by drawing from those very traditions. Spanning the two traditions but nevertheless urging that one is obsolete, their arguments are at best a bridge over troubled waters. Interestingly, their sincerity on this score seems every bit as intense as the ease with which they are willing to sweep away the grandeur of history to make room for the new electronic order and the demands it will make on our time and attention.

It seems clear that the Internet represents a new and unique form of human discourse—certainly a kind of ongoing conversation—and an exchange of information on a scale unprecedented in human history. But, despite our awe at the immediacy of this exchange, can we possibly grant the Internet the same status as the millennium-forged summation of collective knowledge that is Hutchins's Great Conversation?

Why is it that the strident immediacy of the present and future as a temporal impulse to move toward the electronic agora can be so easily conflated with the gravitas of cultural depth and the richness of historical accretion? Is this impulse, in some facile pop-culture sense, simply another "millennial thing"?

Especially interesting is the ironclad certainty of those who would grant the Internet—whose long-term educational value, however promising, still needs to be proven—a status on a par with that of the deep archetypal patterns and traditions of myth and culture. This particular intellectual contagion can be found in some surprising places. Social commentator James Burke, for example, points out that the values of the Great Conversation have been sanctioned by various elites throughout history and that this sense of the specialness of these values must now give way to a more democratic approach, forged by the new electronically enhanced collective mind of humanity. In an interview, Burke observed: "For publishing, what I think we're about to do is move from a culture of scarcity to a culture of abundance. This is why we should put in the deep freeze all of what we used to call the intellectual high ground—great art, great music, and so on."[4]

Burke goes on to acknowledge that standards—presumably what we judge to be of lasting merit versus what we decide is ephemeral and case specific—have hitherto been watched over by relatively small groups, that is, various elites. He then suggests that, as a result of the technological diaspora and the advent of the Internet, we will soon arrive at a condition of civilization in which the standards will disappear: "There will be no standards in the old exclusive sense of the term," Burke contends.

Whether Burke's hypothesis has any validity is certainly an intriguing question. It is, however, a far easier question to ask than to answer. To answer it, one must treat the Internet—a squirming, protoplasmic nexus of informational activity that is never the same from one day to the next—as a conceptual whole of sorts and then project with some measure of certainty the capabilities of its long-term future.

In general, the experience of conversation and dialogue on the Internet is less than overwhelming. Writing in *The New Republic,* Gary Chapman, a researcher at the University of Texas, summarized it this way: "The general quality of the rhetoric on the Internet is discouraging in itself. Even without all the cranks, poseurs, charlatans, fetishists, single-issue monomaniacs, sex-starved lonely hearts, mischievous teenagers, sexists, racists and right-wing haranguers, many participants in unstructured Internet conversations have little of interest to say but a lot of time in which to say it. Goofy opinions and comical disregard for facts are rampant."[5] It is certainly true that the quality of discussions on Usenet and in various online venues ranges from the sublime to the ridiculous, with the latter often getting the upper hand. In fact, the quality of information on Usenet has declined dramatically in the last few years, and across the Internet unhappy customers have made Spam a major concern for companies that provide online services.

While it is easy to point out that Internet discourse is certainly not rife with the wisdom of the ages, the Internet is nevertheless the place where the core values of humanity seem to get hammered out at ground level rather than in the stratospheric soarings of detached thinkers. The questions that have to be decided are whether or not this is a worthy trade-off and to what extent can crystalline shards of truth emerge from the postmodern relativism of what journalist Sander Vanocur has called "a latter-day Tower of Babel."

The hope and the promise are clear: that the critical mass of dialogue will transcend our everyday institutional hypocrisy, the built-in cognitive dissonance of a society (and a government) that says one thing and does another. But more than this, the hope is that the Internet will forge in the white-heat of information long kept compartmentalized, a new compact, that a new view of the world will emerge from the dynamic of human history itself. In forging this compact, the Net's capacity to unleash synergies of human thought long kept in abeyance by the entropy of institutions is paramount. This synergy is the notion behind what *Wired* executive editor Kevin Kelly calls "hive mind."[6]

But must this new collaborative thrust toward the future that the Internet fosters be built at the expense of traditional standards and values? The answer is no, and Burke's slightly too facile abandonment of standards should be uncoupled from the new sense of democratization that publishing on the World Wide Web—at least for now—seems to offer. They are not necessarily correlative. Deep-freeze all great art and music? I don't think so.

The real danger is that the two value systems, old and new, will fail to collide. A collision might at least foster evaluation, dialogue, change; but the concern is that they might just quietly exchange places in the chaos of a transitional society "distracted from distraction," in the words of T. S. Eliot. This exchange can happen only if those who cherish the traditional anchors of myth and history are wary of the new technocrats' increasing sphere of influence and understand the implications of the technocrats' statement that the Internet will someday embody all knowledge. By inference, such a statement must be construed as cultural code for permission to deep-six not only great music and art but literature as well.

What madness is this? Given that all that is of value ultimately rests in some fashion on the continuity of the past and given our increasingly hypnotic attraction to the techno-future, this loss of music, art, and literature would be a huge price for admittance to upscale but largely unknown and uncharted digital realms.

Internet Economics:

The Complex Synergies

of Wealth Creation

Can computer networks generate wealth? This is a difficult question and one that our existing econometric systems seem unable to deal with. To frame the issue properly, we must draw distinctions between real wealth and symbolic wealth. Symbolic wealth is familiar to all of us as the increasingly complicated global monetary exchange systems that enable electronic market trading.

As we transition away from traditional forms of economic power— forms based on property and facilities-ownership—toward the more etherealized concept of wealth as information, the old economic systems of negotiation and measurement fall short of providing the mechanisms to address this change. Furthermore, older economic systems based on the "smokestack" model involved a highly abstract notion of wealth, which has traditionally been defined in a rather limited way.

The concept of gross domestic product (GDP), for example, leaves much to be desired as a method for measuring real wealth. With its inherent focus on symbolic wealth, the GDP excludes many important parameters that should be incorporated in a larger, more expansive definition that also encompasses not only the simple accumulation of capital but also quality of life.

An article by Clifford Cobb, Ted Halstead, and Jonathan Rowe in the October 1995 issue of *The Atlantic Monthly* attempted to redefine the current understanding of GDP and to bring it more in-line with emerging notions of real wealth. The article argued that under the current definition of the GDP not only are many important intangible values that contribute to the quality of life not factored into its calculations but also events and activities that contribute to social deterioration often show up as positive economic elements. One example is economic activity related to the high divorce rate in the United States—economic activity that shows up on the nation's bottom line as positive growth but in reality has negative social consequences.

In many senses, the new information economy is wreaking havoc with our traditional systems of measurement. And since we are in the midst of a massive social transition and reordering of priorities, no easy answers are forthcoming.

The redefinition of wealth will be a slow and gradual process. The good news is that the economic establishment seems to be waking up to the notion that something is awry in its shopworn calculus, especially since the classic economic cycles that defined the "dismal science" for so long seem to have disappeared. In February 1996, the nation's chief economist, Federal Reserve chairman Alan Greenspan, noted the need for change in a speech before the National Governor's Association. Greenspan noted that while "fallout" from rapidly evolving computer and communications technology has created severe economic insecurity in a significant segment of the nation's workforce, this technology ultimately holds the promise of improved living standards. Greenspan also raised some interesting questions, questions that he admitted had no simple answers. He noted, for example, that the nation's GDP is increasingly generated through the use of ideas rather than hard goods, but he also found it puzzling that national output has not shown corresponding gains with the advances in computer technology. Greenspan suggested that the problem might be that "output may not be measured correctly" and noted also that it was simply too early in the game to sort these issues out effectively.

That the widespread application of computers and communications via the Internet is creating a mechanism for wealth generation seems indisputable at this point. One way that computers generate wealth is by creating synergies, a concept originally developed by Buckminster Fuller. There are many ways in which synergies are created. Apart from creating unprecedented efficiencies in the new economic order, computer networking also obviates the deeply embedded, multilayered inefficiencies that have become characteristic of the institutions of modern corporate and government life.

As any management consultant can tell you, the large and ponderously structured corporations that became the mainstay of economic growth and development in the twentieth century are enormously inefficient, moving with all the agility of a dinosaur. These inefficiencies can be traced to a number of factors, but among the most important are parallel critical paths and duplication of function.

When too many layers of management or bureaucracy are present, organizational opportunities become bogged down in a sluggish information flow and are never acted upon or are acted upon too late. The same holds true for the needs of customers, since customer requests often require rapid decision making and a quick commitment of corporate resources. Leaner organizations, made possible by the advent of computers and communications, tend to respond far more swiftly to customer requirements.

The notion of eliminating inefficiencies caused by duplication of function and bureaucratic "whirlpools"—which do not produce a positive and conclusive effect—does not apply only to large corporations. It can be extended to all aspects of our current system of delivery of goods and services, ranging from small and medium-sized businesses to government operations like the U.S. Postal Service to professional services.

One example might be the services provided by attorneys, which include standard functions that are continually repeated. The fairly standardized task of drawing up a will is duplicated many thousands of times throughout the course of the nation's workday by practicing attorneys and their clients. The attorneys who perform this service are compensated in the form of symbolic wealth, but how much real wealth is generated by that activity and how much duplication of effort is involved? This activity has been taken over by software, and the resulting increase in efficiency provides a direct measurable benefit to consumers because the real wealth of the knowledge base involved has become more accessible.

Unfortunately, some of the entrenched professions are in the less-than-desirable position of being antisynergistic; that is, they accumulate their wealth by creating and perpetuating a kind of innocent inefficiency in the delivery of their services. The information age, in this sense, conspires to radically change these professions, as in the case of the Web providing medical information and disseminating it widely. Specialized information thus is no longer the exclusive province of the high priests of any given profession, and the democratization of real wealth (knowledge itself, in this case) is enabled.

Little substantive data convincingly and empirically demonstrates that computer networking—including the force-multiplier effect of the Net—is inherently a wealth-generating mechanism. Most of our information at this point remains anecdotal and is based on simple observation and speculation. (Incidentally, digital culture seems to have an especially keen understanding of the prospects for the synergies of wealth creation, and commentators like Kevin Kelly of *Wired* and Ken Goffman formerly of *Mondo 2000* have shown particular interest in this subject.) But without sophisticated economic measurement systems in place, or at least a better theoretical model with which to proceed, the so-called productivity paradox, as it is known in business circles, will remain a puzzle.

The few studies that have been done to date on computers and wealth creation have been limited to the use of telephone systems in lesser developed countries (LDCs) and have not involved computer networking directly. However, a study performed by the United Nation's International Telecommunications Union (ITU) has shown some interesting results. In 1995, the ITU performed a study, reported in the *World Telecommunications Development Report,* that examined the effect of telecommunications capability on the economic profiles of a number of LDCs. The study tracked teledensity—a measure of the number of phone lines available per capita—against GDP. The resulting analysis, called a Jipp Curve, shows a clear positive correlation between teledensity and wealth generation. We can extrapolate that the effects of the Web and computer networking would likely show a similar degree of correlation. Clearly much more research needs to be done in this area.

If we assume that computers and communications and their capacity for workforce automation can generate new forms of wealth, then the dissonance that emerges from this econometric analysis takes on another dimension altogether and must be recast as a problem of equitable economic distribution and perhaps even, in a more political sense, as a problem of social justice. The intractable and unmanageable aftereffects of the computer-driven global economy have, in fact, become the new focus of our attention as economic interdependency continues to takes its toll on specific world regions and as concerns about global recession continue to hover in the background.

This problem can no longer be solved by any one country: A new economic system will need to be designed to compensate for the somewhat artificial and transitory phenomenon whereby the idea of information-as-wealth shifts the vast majority of computer-related economic gains toward any given society's symbolic analyst community or those who manage the wealth associated with this group.

The distribution of computer-enabled wealth creation is one of the central dilemmas of our times. Currently, we have no viable mechanisms in place to distribute this new wealth equitably because, as we have seen, we have not even adequately redefined wealth either in a societal or economic sense.

The most pressing need is not simply to recast economic models to reflect these new forms of wealth generation but also to make the shift from measuring not just symbolic wealth but real wealth as well. Real wealth for a nation, for example, is semantically tied to the notion of common-weal, a sharing of society's abundance by all its members. Measuring real wealth means looking at quality-of-life issues and recognizing that there are many segments of the workforce that contribute real wealth but are not adequately compensated with symbolic wealth. A cornucopian economy—if that is indeed where we are headed—demands that we abandon the Darwinian zero-sum thinking that nineteenth- and twentieth-century economics has fostered. We also need to consider the possibility that economic instability and even partial collapse might be a necessary precondition for these new systems of distribution.

As other efficiency increases are added to the mix, including large-scale production automation and advancement in collaborative computing made possible by intranets, even more synergies of wealth creation will blossom. In the context of the seismic shifts caused by corporate downsizing and the prevalence of the virtual office, many traditional structures in the workplace—including, as social analyst Jeremy Rifkin argues, the standard of a forty-hour work week—no longer make sense.

The important point here is that it is not a given that we will in fact manage to make this transition successfully, even though much is at stake. We are managing the progress of a new information-age economy with an increasingly irrelevant and rigid smokestack-economy mind-set, in which winner-takes-all thinking still prevails. Worse yet, many of the proponents of digital culture who seem to be in the best position to understand the new economy of abundance seem to be among the least inclined to shift gears toward the notion of a hegemony of real wealth over symbolic wealth. This propensity of cyberculture toward a modified social Darwinism, repackaged in the guise of political libertarianism, remains one of the most disappointing dimensions of that culture's emerging influence over the digital revolution.

The transition that is needed here will not come about easily by any means. Changes of this magnitude will in the short term cause significant pain and dislocation, especially for those at the bottom of the socio-economic ladder, those who may not have the luxury of even a minimal understanding of the significance of the changes.

Thus, the rejoicings of *Wired* and techno-optimists like writer George Gilder are far too premature, especially given the magnitude of what futurist Willis Harman calls the "world macroproblem." Such overwrought enthusiasm will remain inappropriate and even counterproductive to the real work ahead until we convincingly demonstrate that the synergies of wealth creation can solve serious environmental and ecological problems and can creatively and equitably be shifted toward all levels of the newly defined global society. More importantly, we must demonstrate that we can structurally modify our societal thinking and make the shift to a true postscarcity mind-set that surpasses the limitations of the reptilian brain's legacy of boundaried acquisitiveness.

Developing a concept of real wealth is essential to the stewardship of planetary resources at a time of enormous social and political change and daunting complexity. For all its stunning powers of aggregation, communication, and postmodern synthesis, the gift that is the Internet is only the first step in this unquestionably difficult and complicated process.

The Evolutionary Implications

of the Internet

Is the Internet a part of humankind's evolutionary destiny? Certainly, the Net is as much a part of it as any other major technological advance in the twentieth century, but there are compelling arguments to be made that the Internet, as well as electronic media in general, may be playing a much larger role in the evolutionary process than we think. This interdisciplinary field is ripe for exploration and involves an intersection between cognitive science, media studies, telecommunications, and other disciplines at a time when multidisciplinary study is little appreciated. For the sake of argument, we could even toss the emerging field of complex adaptive systems (CAS) into the mix, as well as, I suppose, educational theory, all fields which would undoubtedly be rather uncomfortable with their assigned bedfellows.

Evolution is about adaptation. So is education. The study of CAS is about how systems adapt. Humans adapt both tactically (through education) and strategically (through evolution). I posit here that the Net may not only facilitate adaptation to new ideas but also—far more interestingly—influence genetically encoded cultural and cognitive aspects of human development, what I will call the "evolution of mind."

In trying to address the tantalizing question of evolution, I prefer to include the Net in the larger framework of what has been called the mediasphere, the informational sea of diverse and constantly changing forms of media, the sea that we are immersed in every day. Think of the mediasphere as a series of infinitely regressing feedback loops by which we now—rather casually—view our collective human efforts and their consequences in a global field of awareness. In general, a glaring omission in the current study of the evolution of mind is the mediasphere's role in shaping our intellectual and spiritual destiny. This role is by no means a trivial one.

Although digital culture is highly critical of mass media such as radio and television, most of us can use the media in some productive and positively reinforcing manner. At a minimum, the media allow us to render meaningful judgments on various aspects of contemporary life. And, as Douglas Rushkoff describes in his book *Media Virus,* the most obvious or mundane interpretation of information presented in the broadcasting mode is not necessarily the one that sticks. The "masses" are often far more sophisticated when it comes to making complex discriminations and interpreting the subtler nuances of events than the more elitist media theorists give them credit for.

The essential point here is that values clarification is an ongoing process enabled by the mediasphere's presence in our daily lives. Themes, memes, viruses, and other communications experiments in the art of being human are propagated through the mediasphere with ever greater rapidity and a kind of casualness that can only be described as living theater. Ads rife with pop-culture nuance, situation comedies, Internet newsgroups, talk radio, and all the other facets of the contemporary mediasphere contribute to its variegated richness. To the extent that these entities are cognitively processed in some meaningful sense, we can posit an adaptive response predicated on the necessary reflex of values clarification.

Merely by reflecting the chaos of postmodern reality, the mediasphere resequences expectations and keeps us off balance by moving the goalposts and changing the rules of the game. Media theorist Avital Ronell refers to this process as "scrambling the master codes." I interpret this phrase to mean a kind of resequencing of information that forces us to recontextualize old theories and assumptions about the nature of reality.

The mediasphere also provides benefit by destroying old value systems, which is necessary for adaptive resequencing of information to occur (this resequencing may, interestingly enough, also have a biological counterpart). What many traditional media theorists have overlooked, however, is that this resequencing happens not in an overt way but in a covert or culturally encoded fashion that is independent of the actual content that appears on a television screen or Web site. (I believe that Rushkoff was aiming at this notion in *Media Virus,* although his discussion falls short of the rigor of a formal theorization.) In this sense, the media operate upon us at an unconscious level.

These media feedback loops—and the Net may eventually play a much larger role here—shape consensus reality and define the borders of alternative realities. In the last analysis, they also play a strong but under-appreciated role in the evolution of mind. The sharp differences between the sixties, seventies, eighties, and nineties, each decade with its unique defining characteristics, are in large measure driven by the media's abil-ity to present information as well as by the cascading reaction to that information as compared with the original formulation. (Each feedback loop includes the process of comparison!)

French philosopher Michel Foucault may be right that human nature is as malleable as clay and that we build our own ontologies in the progression of culture through history. If he is, then the mediasphere represents a massive inventory of human experience upon which the process of selection and aggregation is based. The media and the Net may then act in the fashion of a linear particle accelerator, propelling us ever faster through a range of choices that will eventually cluster around a new definition of what is human—in short a quantum leap in the evo-lution of mind.

Explorations of consensus reality undertaken in the sixties suggest that our sense of reality is pieced together from collective individuated perception and is, in this sense, fabricated. Our ethics, social values, and philosophy are, in this view, all negotiated or derived positions. By exten-sion then, the ongoing internal dialogues that are philosophy, religion, and myth are then simply grand thought experiments by which we create new models of reality by seeing what sticks. We then assume that what does stick enjoys a more thorough existential framing or grounding. This point of view, with its obviously postmodern overtones, certainly provides a reasonable explanation for both the whims of science and the fickle his-torical relativism of the dominant paradigm: here today, gone tomorrow.

But are there not also dangers in such notions? To posit an evolu-tionary free hand also means that we must not only face the implicit eth-ical vacuum this freedom creates but also be prepared to reject any postulated sense of archetypal human pattern recognition, such as the great spiritual and mythical patterns that traditionally deepen and broaden human experience. In trying to determine whether computers are part of the problem or part of the solution, we must not overlook the subject of technology-as-resacralization.

There is a wild card in the formulation of an answer to this conundrum: In some deep and unexplained way, the realm of the ancient is tied to our notions of an amorphously envisioned postmodern future. For those who require a larger context to see this process in action, the outlines of this shadow-puzzle are evident in contemporary literature. In Michael Crichton's novel *Congo,* images of high technology subtly resonate with images of our primal ancestry. In the movie *2001,* based on the science-fiction novel by Arthur C. Clarke, the image of the mysterious monolith is juxtaposed against landscapes familiar only to our primitive forebears.

There is an evolutionary linkage here that intrigues and tantalizes but eludes easy explanation. The computer may enable us to envision the future by reinventing the present, but it also might represent a key to unlocking the past and its forgotten knowledge buried under the accretions of sophistication we call civilization and culture. In a way that we cannot fully appreciate now, the technological road that lies ahead may take us on an inexorable journey back to the future.

Notes from Santa Fe:

The Internet as a

Complex Adaptive System

The study of complex adaptive systems (CAS) is an emerging branch of science that has its roots in cybernetics and general systems theory, both of which were developed during the 1960s. The study of CAS, which is related to complexity theory, purports to extrapolate certain fundamental principles of self-organizing natural phenomena into various domains of human activity.

By extrapolating in this way, scientists hope both to understand the "hidden order" of the universe's structural and evolutionary patterns and to apply those principles in a practical way to the design and implementation of areas of endeavor as wide ranging as the development of the next generation of computers, new theories of economic behavior, and theories of business and information systems. Like natural systems, nonbiological systems evolve dynamically and nonlinearly, and complexity theory seeks to identify schema that can explain and predict the development of complex nonbiological systems.

Much of the pioneering work on CAS is being conducted at the Santa Fe Institute in New Mexico, where resident scientists like Stuart Kauffman have been developing the basic principles of complexity theory. Even though the theory is relatively new, applications for it are already being worked on. Business organizations, for example, see wide applicability of CAS theory in predicting the success or failure of various business models. Accordingly, major international consulting firms such as Coopers and Lybrand and McKinsey and Company have visited the institute seeking new ways to apply the theory to the business problems of their clients.

In general, CAS theory does not appear to have been specifically applied to the Internet. In the course of my research for this book, I spoke with Stuart Kauffman about this. When asked whether there were active projects purporting to view telecommunications or the Internet, or both, in this light, he answered no. But he did not see any reason why the Internet, technology's largest and most impressive complex adaptive system, could not become the proper province of CAS theory.

One early attempt to link complexity theory to the field of telecommunications was undertaken by the Aspen Institute, a nonprofit communications-policy think tank based in Washington, D.C. In August 1993, the institute convened the Second Annual Roundtable on Information Technology in Aspen, Colorado. Participants in the conference included John Seely Brown, chief scientist at the Xerox Palo Alto Research Park; computer industry analyst Esther Dyson; Professor Thomas Malone of the Massachusetts Institute of Technology; Judith Hamilton, president and CEO of Dataquest; and Marie-Monique Steckel, president of France Telecom's U.S. operations.

During the conference, one of the key areas explored was the application of complexity theory to both information theory and business systems. The follow-up report on the conference stated: "Many information theorists have enthusiastically embraced complexity theory because it provides a richer, more dynamic model for understanding the actual functioning of the world. Unlike Newtonian physics or neoclassical economics, complexity theory helps explain unique historical contingencies: why one path of development (in an organism, a business, a marketplace, a political system) emerges instead of another, for example."[7]

With roots in both biological science and cybernetics, CAS theory is an interdisciplinary no-man's-land. But a return to unique combinations of various disciplines may be just the ticket for tackling the new terrain. For example, the marriage of media studies, neuroscience, and complexity theory may provide some of the most exciting ground for future study imaginable during the next few years, even though there are few precedents for this alliance and practitioners in these fields might be hard pressed to justify such a marriage.

The field of study here, however, would properly include not only the Internet itself but more importantly the user response to it. With the complexity of the Net mimicking on a large scale the neurocomplexity of the human brain itself, this is a potentially fascinating area of exploration, however elusive it may prove to be in practical terms. The Internet might be described as part technology, part human interaction. To describe it as one or the other is not quite accurate. Unlike other technologies, it does not do anything in the absence of the human mind—in fact, the human mind is the sole source of its viability. Accordingly, the destiny of the Net will be shaped by the interaction of two types of adaptive agents: the systems and software of the Net and its human users. The idea of two adaptive agents resonates well with some of the issues raised in Kevin Kelly's *Out of Control,* one of the defining works of cyberculture. As a card-carrying hyper-technologist, however, Kelly makes a fundamental intellectual error by overemphasizing the technology itself as the vehicle for evolutionary adaptation; in fact, it is our own particular human response to the Net that constitutes the most important dimension (and yet, oddly enough, it remains the most neglected aspect of the ongoing inquiry).

If the Internet is indeed viewed as a complex adaptive system, then the ways in which it might grow, adapt, and evolve are certainly fascinating areas for exploration. Initially, the more salient and recognizable transformations will likely be centered on the areas of scientific research and business development (although these areas are converging somewhat). But can the principles of CAS theory also be applied to the process of government itself, as many of the proponents of digital culture seem to think? The conceptual basis for such a notion has been framed in another report from the Aspen Institute called *The Information Revolution: How New Information Technologies Are Spurring Complex Patterns of Change,* in which the authors suggest that CAS theory could easily be applied to the larger sphere of "the evolution of information technologies, business organizations, markets, and democratic governance."

What all of this suggests is that the dynamic nature of both research and business, as informed and shaped by information technologies and the Internet, will require us to develop new social, organizational, and econometric models to understand their operation. CAS and other theories of complex behavior may or may not ultimately be successful in describing the new patterns. As these models are developed, some measure of vigilance is in order to ensure that a technocratic approach to politics and government does not sidestep the democratic process in an attempt to improve it. Furthermore, as one of the Aspen session participants cautioned, a kind of neo-Darwinism lurks behind these new theories, and as a result they might eventually prove counterproductive.

As the Internet evolves, its human users—we the people—will find ways of adapting to the new technology and, more importantly, adapting the technology to human purposes. Any language that suggests that it is humans that need to do all the adapting is inherently suspect as crypto-technocratic code. For example, consider the following passage from the report titled *The Information Revolution:* "The types of complex adaptive behavior . . . can be seen in at least three important areas: the increasing coordination of the diverse parts of a business organization; the methods by which productivity improvements are pursued; and the human adaptations that managers and employees are making in response to new information technologies." This language sounds innocent enough—we hear it all the time in business circles without blinking an eye. And it could very well be innocent. We do need to adapt to new technology on a day-to-day basis. However, it is the larger sense of cultural adaptation that is the concern here.

Technocrats see humans adapting to the Net, whereas humanists see the Net as a tool in the hands of its users that can be adapted to meet the more critical and pressing needs of society at a time of great social change. There is a fundamental and essential difference between these viewpoints. Thus, attitudes toward the technology itself—how we approach and think about it—are critically important.

Virtual Nightmares

We are drowning in information but starved for knowledge.
—John Naisbitt

Refuse it.
—Sven Birkerts, commenting on
the Internet in *The Gutenberg Elegies*

Painted cakes do not satisfy hunger.
—Proverb

Work, Leisure,

and the

Overthrow of Matter

The ways that new technology can influence our cultural mazeways are myriad and subtle. The invisible tendrils of change tug at our most cherished and comfortable habituations even while we sleep. While it might seem at first glance that the effects of the Internet and associated technologies on business can be neatly partitioned off from their effects on society and culture, the fact is that patterns of commerce are intertwined with the structure of society. We are what we consume, and our lives are shaped every bit as much by how we consume it.

The epistemological method of breaking down a given knowledge base (the Internet, in this case) into microsegments available for purer study has become a crutch, one that we must abandon in order to understand the effects of these new technologies. It is better perhaps to look at these trends from their point of origin: the barely noticeable microevents of daily life, multiplied by millions of individual lives, that are gradually, through the processes of erosion and repetition, being written into our cultural code to the extent that we recognize the sum of these small events as a trend.

Computers and communications technologies are impacting daily life in ways that seem far more important once their effects are tallied in some meaningful sociological sense. And, not surprisingly, behind many benefits lurk negative consequences. Perhaps Emerson's Law of Compensation is at work, or something akin to the balance in Eastern philosophy that is yin and yang. There are few things in this tarnished world that can be described as pure unalloyed good. The trick is to tease out the net result.

Let us consider, for example, marketplace acceleration and time shifting. To an increasing degree, technology allows us—ever more godlike, since we use "commands" to do so—to ignore the traditional boundaries of space and time, to access information, people, places, events, and other forms of virtual otherness at our whim and convenience. One of the often cited benefits of the information age is that it offers "time and distance insensitivity."

Thus, the virtual world has not only "overthrown matter," as technology writer George Gilder puts it, but also conquered the limitations of physics, just as, to a lesser extent, the invention of the telephone did a century ago. But in the construction of these new virtual worlds, why does it so often seem that we are not just tinkering with physical limitations but perhaps engaging in what might be playfully described as a kind of metaphysical cheating, a bargaining that we can only hope is not Faustian in the long run?

In the area of financial trading, the advent of an electronic polity impervious to time and distance means that economically significant events in far-flung locations occur within the seamless continuum of a timeless marketplace that hums along twenty-four hours a day. The entire financial industry is a now a "Store 24" that never closes its doors; the industry is wired to our collective nervous system and is sensitive to the smallest events no matter when and where they occur. Electronic communication has effectively obliterated time zones.

This independence from time and distance has not, however, come without a societal cost. The city that never sleeps has become the planet that never sleeps. This phenomenon has allowed business time, with an urgency fueled by the always hovering promise of financial gain, to invade personal time. Our new and somewhat involuntary immersion in the electronic polity seems to bestow us with the capability to erase the traditional boundaries that have delineated human activities—the very signs and guideposts of cultural comfort and familiarity.

Shifts between various activities have in the past been associated with certain communal and ceremonial dimensions of societal interaction and activity. The increasingly quaint notion of designating one day of the week as a day of rest, for example, represented a kind of fire wall preventing the intrusion of the merely commercial into the aspects of life that are ceremonial or religious or both.

Traditionally, the idea of ceremonial space had both civic and sacred dimensions that were intended to be untouched by commerce. But for evidence that the very structure of society is being radically transformed by the accelerated marketplace, we need only look at the new generation of superstores, patterned after Wal-Mart, that never close their doors. In a kind of societal chain reaction, these stores are simply responding to the needs of individuals whose time is no longer their own. Even the U.S. Postal Service has now begun Sunday operations in many major cities in deference to this "time shortage." And while there are benefits to these conveniences, there are also signs that a kind of backlash to these trends is occurring across broad sectors of society.

The information elite—which does not necessarily correspond completely with the digital elite discussed throughout this book but rather consists of the members of the larger class of knowledge workers that Robert Reich has called "symbolic analysts"—have now learned to conduct their business largely in the virtual realm. In this mode, "business time" is increasingly being broadened to twenty-four hours a day, seven days a week; and this modality is aided and abetted by the asynchronous nature of the Internet and other forms of electronic communications. Images in the media that have been injected into the dumbed-down zeitgeist that is popular culture now advertise a new human modality that is neither work nor leisure but curiously in-between.

While these images appear to promise a new kind of freedom to the upscale, on-the-go professional, they have also forged a new electronic tether to the office, a tether that intrudes further into traditional private and personal space. This tether seems to benefit the corporation and secure the corporation's place in the frenetic new global economy far more than it frees the employee.

A popular and ubiquitous advertisement for paging devices, for example, depicts the tanned and swim-suited body of an executive basking on the shores of a Caribbean island with her paging device ever at the ready in case stock prices suddenly shift. *Nova*, the popular public-television science series (which is biased toward the conventional wisdom in matters of science and technology), produced an episode called "The

Seamless Society." The program's narrative strongly argued that new digital technologies are the avatars of personal liberation rather than, as some critics charge, invisible tethers to a hypermanic and out-of-control global economy. In one segment of the program, the founder of a small but highly successful public relations firm in Sausalito, California, was interviewed. The segment depicted his frequent motorcycle jaunts into the beautiful scenery of the California mountains, where, suitably equipped with a laptop in his saddlebag, he would set up shop on a log or tree stump to keep in touch with the office.

To some, this scenario might represent a kind of utopian fulfillment whereby the commingling of the best of both worlds makes possible an unprecedented personal freedom. The argument is as follows: With the aid of the electronic tether, knowledge workers can truly relax during their vacations. In theory, they are freed from worrying about some random cataclysmic event wiping out their stock portfolios or worrying that some problem with a major project at work will arise.

For others, however, this curious state of half work and half leisure constitutes the very definition of postmodern hell: The individual is never fully in a state of either relaxation or productive professional employment but rather is suspended in the continual tension between the two. Thus, a kind of trade-off emerges: Work becomes more superficially playful as computer games are added to the repertoire of the day's events and as jeans and sneakers become de rigeur; but play also becomes increasingly work-oriented. In fact, the two distinct areas of human activity may be merging into something that is altogether different but shares attributes of both.

Some observers view this phenomenon as clear evidence that a relentless, electronically stimulated marketplace that has forgotten how to address human needs has finally managed to invade what little was left of personal, civic, and sacred spaces in an our already highly secularized society. They argue that in the new 24-7 paradigm (twenty-four hours a day, seven days a week), business seems the clear winner over the individual. Even if the individual manages to temporarily surround his or her frenetic ministrations in the global marketplace with beautiful scenery, that fundamental human experience of standing face-to-face with nature is diminished by the lack of full attention to the experience.

It seems clear that the new utilitarianism has contributed to the increasing secularization of American culture and to the displacement of traditional humanistic and spiritual value systems. In this new society the business of culture is the culture of business. One need only read the crisply rational and highly disciplined views of business-guru-with-pretensions-to-social-critic Peter F. Drucker to get a glimpse of what this unquestionably efficient but soulless and ontologically desolate new society might look like.

Tube Time:

Power Cocooning

for Fun and Profit

The increasing popularity of the Internet suggests that there is a trend toward spending more and more personal time "power cocooning"— using the Internet intensely and nearly continuously. To the extent that one also benefits professionally and economically from doing so, there are further incentives and justifications for this practice. The online world is the realm of symbolic analysts, and participation in this information-rich networked community provides measurable career benefits. Thus, as we use the online world both for leisure and education as well as for professional activities, tube time, time spent sitting motionless in front of a glowing electronic box, continues to mount up.

Spurred by a kind of creeping incrementalism, tube time can eventually constitute the better part of an individual's day. Assume that in the course of a normal day most of us enjoy sixteen waking hours. If we take, say, an hour of "maintenance" time out of those sixteen hours, we end up with fifteen usable hours. It is entirely possible that during the workday, five to six of those hours are spent in front of a computer screen. In the recreational mode, we can add (conservatively) another hour in front of the television to catch up on the news. If two or three more evening hours are added for more television, Web surfing, and other online activities, then the numbers start to look interesting:

> 5 – 6 hours Professional activities
> 1 – 2 hours Television
> 2 – 3 hours Net surfing or additional TV
> Total: 8 – 11 hours of tube time

Keep in mind that these numbers are conservative: They do not take into account individual differences, and they discount special conditions such as Net addiction.

What is interesting in these numbers is the insight they provide into how shockingly close we as a society already are to spending most of our waking hours staring into a computer screen or a television. And all of this is happening before the Internet has matured into something that will presumably be an even more critical part of our daily round of activities. And it seems reasonable to assume that when the Net matures and becomes even more essential for both educational and business purposes, these numbers will only increase, as will the number of individuals who fall into the parameters of this profile. It would not be unusual to see individuals spending twelve to thirteen of the fifteen waking hours in front of a screen, communing (or not) in virtual space.

It is important to see this kind of activity for what it is: sedentary and cerebral. I do not mean to argue that such mental activity might not be valuable or that it does not involve accomplishing significant tasks, enhancing personal education, engaging in interesting discussions, and so on. But consider this: During this activity, the individual is, for the most part, physically immobilized and uniquely disengaged from the physical world.

The possibility that an individual might spend that much time in another "dimension" suggests that we should at least consider the behavior as pathology: a flight from reality into a zone of psychological engagement with its own unique properties. The notion of the body itself being in a kind of suspended animation that mimics physical disability is, of course, an extreme view, one that I am offering as a thought experiment only. But the fascinating relationship between technology and physical disability has been explored by such observers as Avital Ronell and Marshall McLuhan.

Thus, from a purely psychological standpoint, power cocooning could, in theory, be construed as an electronically enhanced fugue state. In this state, individual awareness is diverted toward a field of engagement—the virtual world—that lies metaphysically between the physical world and the pure world of the imagination. It could almost be described as "packaged imagination," although that term would not be accurate insofar as "packaged reality" is involved as well. Virtual worlds may be of one's own creation, they may be provided by someone else, or they may be networked experiences—shared consensual hallucination, to use the language of cyber-influenced science fiction.

In this context, it is interesting to speculate about the timing of the emergence of cyberspace as a phenomenon or, more specifically, as a historical marker. Is it merely coincidental that the virtual world is emerging at a time when significant environmental degradation and breakdowns in the social order are occurring? While the prospect of destroying the planet has existed for years as the result of nuclear armaments, the nineties have yielded the first readily accepted scientific evidence that we may, in fact, be gradually damaging the earth and undermining its habitability.

In 1996, for example, a major United Nations study, which garnered the opinions of a number of respected scientists, concluded that global warming could have potentially significant and severe effects on world health, in no small measure because of major shifts in insect vector populations. Some scientists have predicted that Mexico City will be an environmental wasteland in a few years, with unbreathable air and undrinkable water. As a result of ozone thinning, the chance of contracting melanoma is expected to increase from 1 in 1,500 in 1996 to 1 in 75 by the year 2000. As the reader is undoubtedly aware, both anecdotal and statistical information on this subject is plentiful.

These scenarios are not welcomed by those who are incapable of moving away from a fixed worldview bounded by the conventional wisdom of business as usual. The psyche under siege always seeks defenses. Does cyberspace defend the ego by providing a psychological safe haven from such harsh realities? Is it denial in a box? Is the virtual world, via the mechanism of substitution psychology, an alternative realm? Does the flight to the virtual allow us to escape the truly unthinkable notion that our Enlightenment-based civilizational arrogance has betrayed us, that we have indeed managed to destroy our home in the universe, and that our technological progress is in large measure responsible for this predicament? Could the virtual world be a substitution, however ineffectual, for this enormous loss—a substitution that offers us a new world unencumbered by the intractable problems and unpredictabilities of the physical realm, as well as a comfortable diversion from those realities?

Quality of Information:

The Human

Bandwidth Problem

We often hear about the astonishing amount of information on the Internet and about all the amazing resources it has to offer. I am not disputing this. The information available on the Net is indeed truly impressive. But whenever I am confronted with hype about the Net, I like to perform a simple reality check: Why should the Internet's resources create a sense of wonder any more acute or expansive than the experience of walking into the public library of a moderately large city? Why could not this same surfeit of enthusiasm be applied to libraries in general? After all, they not only contain information from all over the world (like the Internet) but also (unlike the Internet) are the repositories of our culture, our history, and the language-based artifacts of our humanity. They contain classical works and writings that have defined the very nature and spirit of human civilization. Should we not be every bit as impressed and awed by this as we are by the Net? To put it another way, why does the Net somehow undermine this sense of wonderment?

Frankly, the fact that we are not awed by libraries makes me wonder. All too often appreciation for the Internet seems to contain an assumption that therein lies a superior knowledge base that will supersede the old media. (Note that the emphasis is usually on the media themselves and not their content).

While our libraries are being neglected, the Internet is being lionized as the bold and ineluctable future. Funding for schools and libraries is quietly and without adequate debate or discussion flowing away from books and toward technology. But the Internet, for all of its admittedly impressive resources, is far more oriented to the narrowness of the present. It is a mere trickle when juxtaposed against the richness and abundance of history's grand sweep through time. We may live to regret our obsession with the present at the expense of the past.

Proponents of digital culture will argue that the Internet will eventually expand to include many of these other areas—the classics, the humanities, history. The proponents will point out that the knowledge base represented by the Internet is ever growing and changing, and that new types of information are being added all the time. But even if we concede this point, there is another complication, our human inability to use and process all this information.

Let me use a rather exaggerated analogy to make my point. I could place five tons of uranium ore next to my lawnmower and speculate as to how this impressive potential energy source might be used to propel that lawnmower. But of course my lawnmower does not need anywhere near that much energy, and, besides, the uranium would have to be converted into a more useful form.

Can most of us say we have already taken full advantage of the magnificent resources of our public libraries, pondered the classics, learned the lessons of history, savored the intricacies of the research process, or achieved wisdom? Why is it that we suddenly need to have more information more quickly when there is already so much we have not been able to utilize? Looking at the Internet's propensity for information overkill in this light suggests that contagious "information euphoria" is once again at work.

These considerations suggest the need for a new realm of information theory—one that concerns the flow of information and our cognitive capability to absorb and process it—again pointing toward an unmined cross-disciplinary area that lies between media studies and cognitive science. In determining the value of information, whether in electronic or nonelectronic form, we must consider its utilization ratio. Any given base of information resources is only as good as our ability to process it in some reasonable period of time.

Several years ago, an affable gentlemen, Paul Nicholson, worked for *Telecommunications* magazine as executive editor. Nicholson, a physicist by training, once made the observation that the amount of data we humans could process per unit of time—human bandwidth, if you will—was about one gigabit per second. I do not know how he arrived at the figure, and I am sure there are plenty who will disagree with his derivation; but let us assume for the moment that he was correct. By today's standards, one gigabit per second is not a particularly impressive amount of information throughput given the bandwidth capabilities of communications devices available.

What am I working up to here? Simply that the information explosion, the digital revolution, or whatever hyperbolic term you want to apply to the phenomenon will be forever limited by our own relatively narrow human bandwidth. Human beings represent a bottleneck, a place where the flow of information is constricted. Or to put it in another, slightly more cynical way, we are the problem and the digital revolution is not the answer.

Looking at this conundrum as part of the history of computers and communications makes things even more interesting. Several years ago, the first real bottleneck to digital information flow was the personal computer (PC) itself. The computer's own internal communications capability (that is, its backplane) simply was not very fast. This particular problem was solved by the advent of more powerful microprocessors, such as Pentium chips, as well as advanced design techniques, which gave the PC more processing power and more "internal bandwidth." But other bottlenecks still existed, among them the networks that were used to interconnect PCs in either the local or wide area. This so-called fast computer/slow network problem represented the next challenge in moving information around more nimbly and efficiently.

Eventually, the PC evolved to the point that it could handle new processing tasks such as CD-ROM access, some limited video, sophisticated graphics, and other capabilities. But, in general, networks were still chugging along at a very low speed relative to that of the computer backplane.

Thus, there were discontinuities in bandwidth capability: All of that computer-resident processing power came screeching to a halt when it hit either the local area network (LAN) in a corporate environment or the Internet via the standard 64-kbps telephone line available through the public network. During the last several years, particularly in the corpo-rate environment, that problem has been solved by the advent of higher speed LANs operating at 100 Mbps and, in the public network, by higher-speed, high-performance V.90 modems, which operate at 56 kbps. There are other solutions in development, such as DSL-based technology and cable modems.

Now that the PC and network bottlenecks have been greatly allevi-ated, the obvious question is where is the next bottleneck? We have already answered this question: The next bottleneck is the human who interfaces with the network.

No matter how much information is available over the network, and no matter how fast the network is able to deliver that information, there are distinct limits to human cognitive capability in terms of processing speed (and, hence, the amount of information handled). A downside to having the whole world at our fingertips is the crushing weight that this represents and the fact that this huge amount of information must be funneled through a single, narrow access pipeline: the computer screen. As computer and communications capabilities have increased over the years, we have been able to increase the speed with which information is sent through that pipeline to the end user. What we cannot do, however, is increase the speed at which humans can process all of that information—at least not yet.

Information Overload:

A Challenge

for Human Productivity?

Quality of information, information efficiency, and bandwidth bottle-necks are closely related to the notion of information overload—a term that seems to have disappeared from common discourse now that we are in the midst of the digital revolution. (Perhaps it has become impolitic to suggest that one can actually have too much information!)

Information overload was a hot topic of discussion during the seventies and early eighties. It was in the early eighties that John Naisbitt came out with his best-selling book *Megatrends,* in which he wrote perceptively, "We are drowning in information but starved for knowledge." Curiously, now that the Web and the Internet have arrived, few commentators seem interested in exploring how Naisbitt's comment applies to the Internet.

I do not wish to suggest that the information explosion that has taken place in the last twenty years is necessarily a bad thing. What I do wish to point out is that it badly needs to be put into perspective. To do that we must ask a few basic questions. First, is there truly such a thing as too much information? Second, is there an inverse relationship between the quality of information and the quantity of information? Third, what is the relationship between information, knowledge, and wisdom, and how do we distinguish among them in the shifting context that is the information age?

Several years ago, I attended a conference on computer networking sponsored by the Massachusetts Telecommunications Council. The keynote speaker was Robert Kahn, who, as mentioned, was one of the developers of TCP/IP, the basic communications protocol for the Internet. Because of a death in his family, Kahn was unable to present his speech in person. To salvage the situation, the organizers hastily set up a videoconferencing system so that Kahn could at least make his presentation virtually.

During his speech, which was plagued by an unusual number of technical difficulties, Kahn offhandedly made what I considered to be a rather startling confession. It was a notion that I would be willing to bet many of his colleagues had experienced but few would be willing to admit. Kahn was discussing the Internet and the subject of e-mail in general, and he described the huge amount of messages that he typically received upon returning from a trip. How could he possibly cope with this sheer mass of information? Simple, sometimes he just deleted most of the messages! Poof! Bob Kahn can certainly not be faulted for engaging in the practice of information self-defense. He either had to blow those messages away or be held hostage to them in time.

One of the reasons that information overload has become a real problem is the sheer volume of traffic on the Internet. As more and more messages fly around the Net, we have more and more messages to respond to in less and less time.

On the surface, we call this efficiency, lacking another name for it. Enter the hapless e-mail recipient, already pressed for time in the accelerated activity of the new corporate environment. As the number of messages builds, their urgency increases. E-mail a week old requires more attention than postal mail a week old, at least in the conceptual scheme of things. Cycle time diminishes as more and more messaging shifts from the "snail-mail" mode to e-mail and information begins to dart around the globe with mercurial swiftness.

The planetary metabolism becomes stoked by this information flow; and this increased metabolism, in turn, speeds up all of our transactions, both business and personal, and curtails the amount of time for deliberation and reflection. In this scheme of things, quick reaction time and rapid cognition—already distinguishing traits of Western culture—are validated as important traits. By the same token, careful deliberation easily becomes devalued because it gets in the way of progress.

As the former editor-in-chief of a magazine that covers one of the world's fastest markets, I could cite my own experience with information

----

overload. Did e-mail make the information-intensive job of putting out a monthly magazine easier or harder? The answer is, without a doubt, harder, at least in the short term. The online environment only increased the magazine's already daunting information flow. In addition to faxes, regular mail, personal visits, conferences, seminars, trade shows, and phone conversations, a huge amount of external e-mail was delivered with predictable and merciless efficiency via the magazine's multiple mailboxes.

If all of that information had been converted to electronic form, would the job of sifting through it to determine what was most useful to our readers have been any easier? From my own standpoint, the answer is a resounding no, and the "one person/one tube" information bottleneck theory needs to be invoked here.

I will now make the heretical statement that, in fact, it is very hard to improve upon a stack of papers for efficient information access. I can riffle through papers in seconds, as fast as or faster than I could access the same material were it stored on my laptop (which is, of course, not possible, at least so far). More importantly, I can create a visually-based conceptual "architecture" for the various types of raw information—and I can assure you there was a considerable amount at any given time of the month—stored in my office. Seeing the various piles of material arrayed in physical space (in stacked trays and other low-tech devices) also helped me to navigate quickly. Compacting all of that material into the confined space of the PC would only create a narrow information bottleneck, that is, the computer screen and the person reading from it (me). The PC ultimately proves to be an inefficient tool for retrieving information. Eventually we may be able to create three-dimensional flat-panel wall displays to duplicate this function, but until this happens getting all of one's information via a computer screen still remains a self-limiting approach.

One small disclaimer to the foregoing: I do not wish to claim that all editors' habits are the same as mine or that my experience holds true for everyone. But as information flow has increased over the years, using the same basic set of resources, I—like other editors and information professionals—have had to face the challenge of accommodating and processing this increased flow. By and large, the solution has been to rely not on more technology but in some ways on less. I call the techniques with which people address this increased information flow "strategies of accommodation." In one sense, increased information is the problem, and our job, as symbolic analysts, is to find the solution.

Here again the notion of coevolution between adaptive agents comes in, an idea derived from complexity theory. Adaptation to a computer-based flow of information needs to occur on a cognitive level. Dealing with this information on a daily basis involves altering personal habits, adjusting one's personal configuration of hardware and software options, and using a lot of clever cognitive tricks to create information arrays in visual space.

Surviving this onslaught of information overload (barely!) by adopting these strategies has led me to the conclusion that we are placing far too much emphasis on the development of the technology itself and far too little emphasis on the human factors involved. This is one of the biggest criticisms that can be levied against organizations like the Media Lab, MIT's computer-technology think tank, where the emphasis always seems to end up on the technology rather than on appropriate human use of that technology. If these powerful new tools are to evolve in new evolutionary and transformational directions, the emphasis will have to change dramatically over the next decade. The bad news: We are using twentieth-first-century tools with a twentieth-century mind-set.

One interesting footnote here pertains to the sustained impact of this increased information flow on human cognitive processes. Up to now, I have, with a certain amount of deliberate coyness, cited the current bottleneck as the human information processor. But a rather fascinating question emerges: Can we increase our ability to process information; that is, does the phenomenon of information overload in any conceivable way train us to process information more quickly and efficiently? On the basis of personal experience and informed conjecture, I do believe that our processing capabilities can be altered, can indeed even be reconfigured and accelerated so that we can adapt to these new forms of information presentation.

Whether increased human efficiency is inherently desirable is another question entirely. The link between hyperactivity in children and the use of computers or electronic games like Nintendo is ripe for exploration. So is an increase in the ability of an electronically nurtured younger generation to engage in multitasking—the simultaneous reception of multiple streams of information—in much the same way that a computer does. For example, we work on a computer while talking on the telephone, and teenagers can watch television while doing homework.

There are many avenues to explore in this arena. Does multitasking diminish the quality of information gained from each activity? Is multitasking a cognitive enhancement or a semischizoid inability to focus? Adaptation to computer-driven regimes may somehow represent a new chapter in the evolution of the mind. But great care is in order: Although the notion of human adaptation to the machine, however sophisticated that machine may be, is cloaked in digital culture's tantalizing metalanguage of transhuman performance, that adaptation might differ only in degree from tethering a worldview to the clock, the reigning symbol of the supposedly discarded Newtonian universe. Our dalliance with the machine may make us more than human or less than human. It may lead to some new quasi-insectile form of human organization—the mercilessly efficient "hive mind" of Kevin Kelly—or, as some cyberculture thinkers suggest, it may force us to redefine the term "human" altogether.

There does seem to be a link between quality and quantity of information. If accelerated information flow forces us to process information less thoughtfully and to become more reactive and superficial, then "hive mind" will have proved a chimera. Information may be exchanged with ever increasing rapidity, but we should not ignore the possibility that the trade-off, the demise of deliberation and thoughtful analysis, might severely degrade the overall quality of information. In fact, in other forms of electronic media, such as television, we have every reason to suspect that this process is already taking place. Witness, for example, how the need to deliver television news to viewers before the competition—the advent of the twenty-four-hour news cycle—has diminished carefully crafted reportage and encouraged the reporting of the sensational rather than the substantive.

Information for information's sake—a value touted by Nicholas Negroponte in his book *Being Digital,* a value increasingly accepted without scrutiny—is an illusion. Information should not be a value in and of itself. Our society, seemingly afflicted with a new kind of information-age amnesia, seems to be forgetting to ask the important questions. What is the nature of the information? What is it source? How reliable is that source? What will the information be used for? Who validates it? Who disagrees with its validity? What is its level of quality?

In thinking through these questions while avoiding the more troublesome seductions of the digital revolution, we can make useful and nuanced distinctions that center on the issue of quality of information and on the differences between information, knowledge, and wisdom. In the last analysis, information is not the holy grail at all. Information is not, as the conventional wisdom is so fond of suggesting, the end product but the raw material for knowledge, which is the ability to interpret information in meaningful and useful patterns.

Once information is garnered, rational thought, reflective analysis, intuition, and other basic skills obtained through a classical education can be brought to bear. The vast amounts of electronic information on the Net and elsewhere, while important and useful, tend to place value on the wrong element of the educational process: data collection. In reality, data collection is only a starting point. It is what is done with the data that is truly important.

When information is converted to knowledge, the true alchemy of the educational system and the very essence of human cognition are reflected. If we overemphasize information for its purely utilitarian value and undermine deliberation and reflection, both of which can connect us to deeper value systems, the process of information conversion will be subverted. Perhaps we might consider the possibility that it is this cognitive reorganizing of information that makes us more deeply human and represents the crowning achievement of our evolution.

The Electronic Agora

and the

Death of History

Information overload is not limited just to the professional sphere and the realm of symbolic analysts. Powerful forces in the technological diaspora are creating an unprecedented diffusion of information throughout society. We as a culture are being force-fed huge amounts of information as it propagates through the mediasphere—a phenomenon that might be called "information bombing."

As a result, our collective attention span is being compromised by the need to keep up. Society itself is plagued by a kind of attention deficit disorder writ large. With so much information about the present constantly flowing across the cultural radar screen, it becomes increasingly difficult to keep up with the present, imperative to try to anticipate the future, and discouragingly difficult to maintain a reasonable focus on the past.

In this age of rapid change, our increasingly fragile sense of history is being sacrificed to the present and the future. For those who wish for a technocratic and market-driven future, cutting the cultural moorings that might interfere with its establishment is certainly a desirable goal. Undoubtedly, the roots of the past can easily trip up the hasty stampede toward digital futurism, however ill-defined that future may be. Some might see religion and history, art and philosophy as speed bumps preventing or slowing the ineluctable march toward a new economic and political order in which electronically distributed information is not only the centerpiece but the raison d'être. A brand new library that opened in San Francisco in 1996 and offered many computer terminals but few books is testimony to the fact that there are some who are only too willing to cut loose these deep roots to the past. (It seems almost unnecessary to add that the library was funded by large contributions from leading-edge computer companies like Microsoft.)

The new technocrats—and I place the members of digital culture squarely in this camp, even though the two are not necessarily synonymous—are interested in nothing less ambitious than reinventing society in their own image. To a large extent, they see certain anchors of traditional culture as standing in the way of their vision of a new society.

I can say without exaggeration that there are few established traditional sources of education and learning, cultural guideposts, and political mechanisms that the new technocrats are not willing to sweep aside to fulfill their rather narrow vision of the future. When *Wired* magazine promotes the idea of the Net as the basis for a new political party, few people really understand just how radical this notion is.

Among the proponents of digital culture, Stewart Brand and Kevin Kelly embody this tabula rasa approach to futurism. In their thinking, history is just another inconvenient stumbling block on the way to a new order, a reinventing of society made possible by digital technology. Both Brand and Kelly pass the technocrat test hands down because they see technology, society, and culture as blurringly confluent. Writing in *Wired*, Brand has, in fact, celebrated the future day when all of our communications will have vanishingly small half-lives and will come and go as if they were "written on the wind."

In this mind-set, impermanence and immediacy become values to be cultivated. While such values might be easily lionized in, for example, the Balinese culture—where the barriers between thought, word, and action are very low indeed—mixing this sense of permanent impermanence, which is based on the vagaries of the information flow of the moment, with the complexities and volatilities of modern culture is bound to be problematic. Witness what happens to children when their lives are not properly anchored in the basic values of community and society but are instead allowed to be shaped by popular culture's meme of the week.

Both Brand and Kelly are also supportive of the notion of a "third culture," a displacement of the ancien régime of literature and the arts as the pinnacle of human achievement with a new locus of value centered on the real-time explorations of science and technology (no matter that the content of scientific theories and admonishments are clearly as arbitrary as cultural values themselves). John Brockman, longtime literary agent for the *Whole Earth Catalog* and other *Whole Earth* projects had this to say about the third culture in an issue of *Wired:*

> The literary culture I talk about is pretty well finished.... It was an establishment that dictated fashionable discourse and prided itself on its indifference to science. It favored opinions and ideology over empirical testing of ideas—commentary spiraling on commentary. As a cultural force, it's a dead end.

Brockman goes on to suggest that we need to build a new culture to replace this defunct and outdated literary culture:

> What we've lacked is an intellectual culture able to transform its own premises as fast as our technologies are transforming us. The only place you're going to find that is in sciences where empiricism and epistemology collide, and everything becomes different. That synergy exists, for example, in the work of biologists Richard Dawkins and Stephen Jay Gould; physicists Roger Penrose, Stephen Hawking, and Freeman Dyson, astronomer Sir Martin Rees; and computer scientists Danny Hillis and Marvin Minsky.[1]

While the work of many of these individuals is indeed at the cutting edge of science, by what process of logic should we assume that it is their work and worldview that should form the basis for culture itself? Furthermore, the sense of humanism and even ethical grounding of some of the scientists named might be considered questionable. Marvin Minsky, for example, is an MIT scientist who fostered the notion of the human body as "meat"; his principal vision of the future involves the bizarre notion of downloading human consciousness into the permanent and immortal physical repository of the computer.

Should we also be encouraged to jettison thousands of years of literary and cultural tradition—Shakespeare, Yeats, Plato, Aristotle, and all the other great minds our civilization has brought forth—in favor of

Minsky's fifties-style mad science and the warmed-over scientific materialism being offered as new and cutting edge by Brand, Kelly, and Brockman? A culture built upon the ideas of Marvin Minsky would be a strange one indeed.

When the emergent worldview of digital culture is properly deconstructed, understanding why there is so little resistance to the process of conversion from books to information in electronic format becomes a lot easier. Books, after all, are historical in their focus and represent our best cultural anchors to the richness of the historical narrative of ideas and events.

Unfortunately, history itself is highly vulnerable under the rubric of this new technological revisionism for some very practical reasons. When the "great transition" (the shift of most knowledge bases from traditional to electronic media) occurs, some material will make it into electronic format and some will not. As Clifford Stoll points out in *Silicon Snake Oil*, what will most likely not make the transition is books—millions and millions of them, each representing the real "embodiment of human knowledge" across a wide variety of categories and disciplines.

Despite a few noble attempts to convert books into electronic format, such as the Gutenberg Project, the fact is that because of sheer logistical difficulty, this process is not likely to occur on a wide scale. Nor is the experience of reading from a computer screen ever likely to approach that of reading a book. The cost of our new obsessive focus on the present and its unrelenting information stream just might be our deeper connections to the past and its rich diversity of knowledge.

Interiority:

Our Most Precious

Natural Resource

A nod, a wink, a glimmer in the eye, the beveled edge of an intellectual's keen awareness, a friend or loved one's priceless gesture, a child's emotional shorthand for pain and disappointment—these are the true currency of human awareness. But awareness itself—that sense of the larger universe in which we collectively dwell—might inadvertently become a casualty of the electronic age.

Closely related to the death of history, the shortening of attention spans, and the increasing tendency, electronically fostered, toward reaction rather than deliberation is the threat that the new electronic polity poses to interiority. Interiority—our ability to step back from people, places, and events and reflect on a given situation or condition—is perhaps what delineates important aspects of our humanity and separates our actions from the split-second reactions of the insect hive or the thundering herd.

In *Silicon Snake Oil,* Clifford Stoll talks about how e-mail, with its insistent demand for immediate response, works against the process of thoughtful composition. He also suggests that the widespread use of word-processing software has changed the very essence of the act of writing. Stoll argues that both reflection itself and the formation of more fully coherent expression in committing thought to paper have been negatively affected.

There are many other examples of how contemplation and reflection have been short-circuited. Neither the Internet nor computers and communications are solely responsible for this diminishment of interiority; rather, it can be seen as a function of the intrusiveness of electronic media, including television, radio, and other forms of communication, in contemporary life, the incessant bleating of the mediasphere.

In this context, we might look at the mediasphere as the externalization of thought. When I say "thought," however, I mean not pure or original thinking as we know it but rather the by-product of our collective thinking or processing of values, averaged out and sanitized for our protection (that is, the conventional wisdom).

If nothing else, the media can be counted on to reinforce this ad nauseam, ad infinitum, thus sparing us the painful task of true values clarification. In the process of externalization, individuals are relieved of the existentially dreadful pressure to make thoughtful decisions by a shaping of norms inherent in the values presented. This is why media values are so confusing to children and why adults are so confused by children's response to the media and the facile adoption of its codes. The more the electronic media externalize thought, the less we have to delve into our own interiority to produce it.

Media saturation is a fact of contemporary life. It fosters a need to react. We now live much of our lives quasi-vicariously through the ever present lens of the mediasphere, the accelerated gestalt of which creates a sense of urgency that is sometimes merited and sometimes not. Reaction versus reflection and deliberation, therefore, is a central motif in considering the sociocultural effects of the paradigm shift from print media to electronic media. But it is also an important and unpleasant side effect of contemporary culture that is frequently overlooked.

The theft of interiority is sometimes based on an undermining of our sense of reality. The dynamic is a subtle one. In the mediasphere, real events are increasingly placed on equal footing with media events. (This gets to the notion of the simulacrum as discussed by postmodern thinkers like Jean Baudrillard.) Media descriptions of events are thus prone eventually to become more real than the events themselves.

There is a fascinating corollary to this: Some real events become media events, but not all media events are based on real events, even though they may commonly be interpreted as such. In this context, the sense of reaction generated by electronic media becomes an extremely important issue. The incessant barrage of the mediasphere speeds up information reception as well as cognition. As a result, society and culture become stimulated and superheated by chain reactions—collisions, if you will of events and information—until, as in a nuclear reactor, critical mass is reached.

By means of a complex series of feedback loops, the mediasphere creates a kind of hall of mirrors in the collective psyche. We begin to see ourselves as a society acting out, both onstage and off, the megadramas of the age. The impeachment trial of President Clinton, for example, showed this hall-of-mirrors effect in all its glory, with comment upon comment ricocheting and colliding with new and old information, changing the shape of events, and pulling the viewer with increasing force into the immediacy of the moment.

The media's increasing self-boosterism and the way that they like to study and analyze themselves suggest that the observed and the observer may have, in Heisenberg's terms, not only affected each other but become fused. There is little room left for anything else. We think we are in a free and open universe of independent events, but the reality is the opposite: We are on a Moebius strip. As we roller coaster around the strip, the media are only too happy to increase the speed here and there to augment the thrill. In the real world (and not the simulacrum that is the mediasphere), we have little time for reflection as we barrel down a steep hill at eighty or ninety miles an hour. Similarly, in the constantly accelerating mediasphere, quick-cut neural stimulation is what sells newspapers.

This stimulation has now become a value in and of itself, a value that is deeply counterproductive to the processes of reflection and contemplation. It may be one of the causes for the ongoing difficulties of our educational system, and, like a drug, it requires increasing doses to maintain its cognitive potency.

We can only speculate about the degree to which both the online experience and other forms of electronic communication contribute to the theft of interiority. It seems clear that there is some sort of linkage between the neural stimulation provided by electronic media—be it MTV, the Internet, or Tetris on a Gameboy—and the increasing phenomenon of hyperactivity and general diminishment of attention span in children and, sad to say, adults and society in general.

These activities appear to erode interiority while at the same time speeding up "neural metabolism." They may result in cumulative negative effects, especially when they are used as replacements for human interaction (that is, as electronic baby-sitters) or as substitutes for the complex neocortical adventuring that young children need for proper development.

The gradual erosion of interiority also has significant implications in the larger societal context of political action and democratic discourse. As the reactive mode becomes more prevalent, the deliberative aspects of due process are diminished. The resulting ad hoc jury-rigging was evident to all during the Clinton impeachment trial.

When this tendency is coupled with accelerated social change, both the electorate and the elected officials find it increasingly difficult to keep up with changes taking place at all levels of society, to absorb them, and to then make intelligent assessments of their impacts. As our democratic institutions become less responsive, reactionary schemes are more likely to take root as quick fixes for problems that seem overwhelming in their complexity and their rapidity of onset.

Thus, the spiral continues, aided and abetted by the electronic media and diminished by the lack of deliberation. The very fact that we call some of our governmental institutions "deliberative bodies" clearly suggests that their function is to ponder the important issues of the day by affording them the proper degree of consideration. As the processes and interactions of our society continue to be accelerated by electronic communication (and they will) and as our problems become more complex, chaotic, and technological in nature (and they will), these issues will become even more egregious.

The dangerous trend toward reaction and away from deliberation lurks beneath the concept of electronic democracy, a shibboleth of digital culture. Superficially, at least, electronic democracy has a certain amount of pseudopopulist appeal. After all, Americans tend to distrust any kind of mediation or third-party representation to begin with. In this regard, comparisons have been made between this distrust and the Protestant impulse to enjoy a direct, unmediated relationship with the Creator.

As appealing as the idea of electronic democracy is, however, close inspection of the argument in favor of it reveals serious chinks in its armor. In the final analysis, a scheme of highly interactive mass communication that fosters reaction over reflection and deliberation may be as inimical to the democratic ideal as the overbureaucratization that has characterized the process of government self-aggrandizement during the last forty years or so.

It is helpful in this context to remember that direct representative government need not be electronic in nature. There have been many experiments in recent years involving voter referenda, especially in California. These experiments bear out concerns that electronic democracy might be considerably less attractive in actual practice.

In the November 1994 issue of *Harper's Magazine*, Peter Schrag analyzed the serious failure of experiments with participatory referenda in California. Schrag contended that rather than reflecting the will of the people, these referenda served only to hopelessly disconnect the means of government from its ends and to create terrible discontinuities in the governing process. It appears that these efforts have shown that making decisions by consensus—whether electronically enabled or not—by no means guarantees that the decisions will be sound.

The technological intrusiveness of electronic media and its mind-numbing assault on interiority seem at this point unstoppable. Stores now sport preprogrammed sales pitches that bleat at customers while they shop. When put on hold by a company that offers shoddy and unresponsive customer service, consumers are treated to packaged dialogue offering earnest assurances in silky tones: "Your call is important to us."

Television screens are now de rigeur in the waiting rooms of establishments from auto repair shops to restaurants to eyeglass emporiums to airports. Many major airport terminals across the country have hung large-screen television sets from the ceiling in their waiting areas, and travelers who wish to read a book or newspaper, do some people watching, or simply sit in silence have to actively seek out a place where they cannot see or hear the broadcast—often not an easy task. And, of course, thanks to Chris Whittle's Channel One and the Edison Project, television and its commercial messages have now invaded the classroom.

The insistent demands of the mediasphere and its often specious sense of urgency target all the untouched and unprotected spaces of human awareness in a relentless effort to pull all of us onto common ground: the all-encompassing sameness of the electronic polity and the externalization of thought. The mediasphere beckons with real events and pseudoevents positioned for an optimal sense of urgency by an increasingly sophisticated array of techniques leveraged by media handlers.

Why not simply opt out and pull the plug? Go ahead if you dare. To disconnect from the mediasphere is to risk the contemporary equivalent of excommunication from the Catholic Church in medieval times. Disconnection from the real-time power surge of the electronic polity means living outside the defined boundaries of what it means to be a vital, aware planetary citizen, at least according to the media's conventional wisdom (and there is indeed much truth to that assertion).

When the first O. J. Simpson trial was in full swing, we got a slightly disturbing glimpse of a brand new phenomenon that I call "neo-Puritan electronic voyeurism." This magnetizing event gave us a fleeting glimpse of how the mediasphere might look at some future point when its ability to evoke a riveting electronic spectacle has been perfected and society's inhibitions about seeing the private-as-public have been further eroded. The trial gripped the attention of many Americans, some of whom jokingly referred to their need for a daily fix. Despite the jokes, alarming signs of true addiction were visible in their apologetic smiles.

The public spectacle that the trial represented illustrated the power of a real event to become larger than life and to take on an almost hypnotic authority when powerfully presented to a mass audience. The phenomenon led one prominent psychologist to comment that some television viewers overidentified with some of the players in this teledrama, such as prosecutor Marcia Clark. Such overidentification suggests the possibility that the media can induce a special kind of pathology and dependency in those who are psychologically susceptible. But, more significantly, it also seems to represent an excellent example of what can happen when media

invasiveness diminishes interiority and self-awareness. Into the abhor-
rent vacuum that results, hapless wanderers in the mediasphere rush to get
their special dose of realities otherwise denied, while the fragile boundaries
of self are quickly overwhelmed by the rush of tantalizing but useless new
information.

Electronic Mediation

and Technological

Dependency

The speed with which we are rushing toward new forms of social control via information technology is startling. Not all of the changes are cause for concern, but the general lack of commentary on the rapidity with which they are being adopted is certainly bothersome.

From the electronically based national identity cards once proposed by President Clinton to electronic tethers now commonly used for prisoners to biometric security to geosynchronous satellites that monitor railroad tracks for signs of vandalism, the advent of new and increasingly intrusive electronic watchdog technologies is propelling us into a future that has been hinted at in popular culture but seen as unlikely by most of us. But these trends are real—and they will deeply affect society and culture well into the twenty-first century.

When viewed in the context of popular culture via science fiction warnings and other vehicles, these trends indeed seem troublesome. But how concerned should we be? In the watchdog technologies cited above, the computer mediates relationships between human beings. This impressively designed but potentially culturally intrusive technology is being subtly but unmistakably positioned to displace varying kinds of human judgment via the process of mediation.

In the book *Technopoly,* Neil Postman discusses how the value system of the technocrat places more faith in the cybernetic measurement of life experience than in life experience directly encountered. This modality is increasingly pervasive in areas like health care, where treatment based on human interaction with patients and on experiential diagnosis is no longer the norm. Patients are subjected to scanning devices and a battery of technology-based tests that objectively inform the medical staff about the patient's condition—ample evidence that the technocratic mind-set is dependent on an atavistic but tenacious scientific materialism.

The technologizing of health care has been a large factor in the increase in medical costs and the decrease in patient satisfaction, and it is a major causative factor in the so-called health-care crisis. In a vivid illustration of the future role of the computer as mediating device and a warning of a deeper kind of alienation ahead, some health-care providers are now considering the use of videoconferencing systems to conduct the initial screening for incoming patients so that medical staff members can avoid direct patient contact (or so I was informed by a company that makes such devices). Companies like Japan's Fujitsu are selling Asynchronous Transfer Mode (ATM) switches to health-care providers with Switched Virtual Circuit capability, a technology that will allow these kinds of arrangements to be implemented easily. These companies are even marketing the switch specifically on the basis of this capability.

From the standpoint of a humanist (or, one could argue, a medical ethicist), the loss of human intersubjectivity in these kinds of scenarios is stunning. Study after study has shown that there are critically important dimensions to health care that have far more to do with the relationship of patient to doctor and patient to medical staff than with the functioning of diagnostic equipment. That we are considering replacing these relationships with the prowess of machines, however sophisticated, is an unfortunate trend.

In this new technocratic way of evaluating reality, nothing is real unless it is somehow validated by computer processing or electronic measurement. We are increasingly encouraged to ignore the direct evidence of our five senses and steered toward the notion that the objective evaluation that can be provided by a computer system is superior to its traditional counterpart: simple human observation and experience.

In the larger sense, it seems clear that when trust between human beings is diminished, mechanical means of verification becomes a necessary (but undesirable) replacement for what was formerly a purely human transaction. Ironically enough, it is the power of the computer itself to forge documents and create simulacra of various types that is in no small measure responsible for the breakdown in societal mechanisms to validate identity and perform other essential functions of business and government. The computer becomes at once part of the problem and part of the solution, and this dynamic continues to feed upon itself, resulting in greater and greater levels of technicization in which the technically sophisticated win and the average citizen often loses.

The Psychopathology

of Online Life

The use of the computer as a means of mediating human relationships raises the basic issue of alienation, a word that seems to have been quietly dropped from our contemporary lexicon. For that matter, the study of sociology itself seems oddly deemphasized—take a look at the sociology section in any major bookstore and you can get a vivid sense of this intellectual attrition. Sociological texts written in the fifties and sixties were very concerned about alienation in modern life. Why has that concern evaporated, and why are hard questions about this aspect of the quality of life being asked so infrequently?

We need to consider the extent to which new technologies engender a kind of commodification of human relationships in a new marketplace, a commodification not only of consumer goods but of displaced systems of value. In addressing the issue of social alienation, we also need to ask whether there is a deeper social pathology at work in our attraction to these electronic modalities, as difficult or counterintuitive as this process might be.

This exploration necessarily involves probing into the darker side of American cultural mores and habits. For example, there have been many critiques in the sociological literature over the last several decades dealing with the pervasiveness of isolation and loneliness in American culture, critiques ranging from the best-selling treatment by Christopher Lasch, *The Culture of Narcissism,* to Philip Slater's *The Pursuit of Loneliness.* The latter focused on the rootlessness of Americans, the ease with which we sever ties of community in the service of what are considered to be the overriding goals of progress and self-fulfillment.

An interesting question emerges: Is the cultural fascination with cyberspace a manifestation of longing for community in American culture? And, if so, does cyberspace represent a cure or merely a pathology-tinged substitute for what has been deliberately or inadvertently shunted aside?

Determining the extent to which the digital revolution is either symptomatic or causative with respect to these patterns in American society is a complex and interesting problem. At this juncture, there seem to be no clear-cut answers. However, the optimal starting point lies in asking the right questions and being willing to openly and objectively tackle the negative aspects of the information revolution. In this context, it is helpful to summon the thinking of maverick psychologist and social observer Ronald Laing, who in books like *The Politics of Experience* asked us to consider the possibility that societies can manifest pathologies just as individuals can.

There are several interesting avenues worth exploring here. The first is fascinating but inconclusive: Many of the major advances in telecommunications have been achieved by inventors with some physical or mental disability. Although many would consider her work to be on the fringes of conventional academia, Professor Avital Ronell has written extensively about Alexander Graham Bell's deafness in her fascinating but somewhat impenetrable book *The Telephone Book: Telephones, Schizophrenia, and Free Speech*. Far less known in the mythos of cyberculture is that Vinton Cerf, considered by many to be the father of the Internet, is partially deaf and that Nicholas Negroponte has a reading disability. Negroponte, in fact, has openly professed his dislike for both reading and writing and begins his book *Being Digital* with the statement: "Being dyslexic, I don't like to read."

This admittedly anecdotal information suggests a kind of overcompensation syndrome at work in the very invention of the technologies in question and raises the specter of substitution psychology. It seems that the virtual worlds being created are construed by their creators, at least to some degree, as substitutions for aspects of reality in the natural and physical world. Intriguing here are the possible psychological roots of the notion of substitution. Is the fugue state that constitutes the online experience in any way analogous to mental illness or any other condition in which people seek to avoid true reality by replacing it with a more controllable or self-manufactured reality?

The phenomenon known among cyberspace practitioners as online addiction should arouse our suspicion. What is also telltale (and disturbing) is evidence that the online world is being given an ontological status that is more or less equivalent to that of the physical and natural world, as if there could possibly be any real competition between the two. There are seasoned online habitués (addicts?) who will quite casually proclaim such equivalence. On the WELL, for example, I have seen participants make this equation without hesitation while offering a spirited explanation as to why spending one's time in the virtual world is superior to grappling with the messy complexities of "meat space."

Are such notions pathological? Or are they simply a wistful desire for some ordinarily unachievable Platonic purity? If the former, then is it a pathology of individuals or of society? Given that our planet is rife with enough metaproblems to push many of us toward either denial or Gurdjieffian somnolence, we can wonder if there is not some kind of escapism at work in the quest for virtual nirvana.

Environmental degradation, for example, is certainly a cause for great concern, but it is clearly not something that our society is dealing with at a rational, conscious level. Given that global warming and massive climate change are distressing possibilities, the retreat into cyberspace can be construed as a kind of flight from reality and a quest for a purer, more rarefied world.

This quest is a wholly understandable, fundamentally human one, deeply rooted in the impulse toward ascent experience and the longing for the transcendent. The universal human longing for an optimized plane of existence is of course the very bedrock of the spiritual quest. But falsely repositioning this quest in the admittedly enticing realm of the virtual world, digital culture appears to be making a kind of category mistake, wherein something with the appearance of the transcendent other has been substituted for the truly transcendent.

It is an understandable mistake, however. Given the quality of etherealization discussed elsewhere, the virtual world indeed appears to have certain things in common with the realm of spirituality. In an increasingly secularized society looking where it probably should not for forms of spiritual reaffirmation, these superficial commonalities are sufficient to mislead digital culture into thinking that the massive global nexus of computers has some real rather than metaphorical connection to genuine spiritual or mystical experience.

Such confusion is evident in the remarks of cyberculture maven William Gibson in a *Time* magazine cover story exploring the link between computers and spirituality. Here Gibson notes, "It seems as though the Net itself has become conscious. . . . it may regard itself as God. And it may be God on its own terms."[2]

If we continue to explore the realm of metaphor, we find that it is, of course, possible that the Internet does somehow symbolize a new emergent world consciousness in the millennium that lies ahead. It is even possible to suggest, as Douglas Rushkoff does, that the two phenomena are somehow related (he ties the notion of expanding digital communications across the planet to the Gaia hypothesis).

But it is also possible to argue that precisely the opposite is true: that the dehumanizing effects of mediated experience and the loss of intersubjectivity caused by the computer are hardly in the realm of the spiritual. As preposterous as it might seem, many Christian groups believe the computer to be the Beast long foretold in biblical prophecy. And any student of spirituality can see that the grotesque neomedieval techno-phantasmagoria that characterizes the world of *Mondo 2000* readers is quasi-demonic in form and likeness.

Clearly, some reality testing is in order. The late Chogyam Trungpa, the spiritual leader who wrote *Cutting through Spiritual Materialism,* explores the Buddhist notion of sanity. Like many of Trungpa's concepts, sanity is an elegant and subtle notion that has much to do with accepting the conditions of one's own constellation of particulars with a certain grace and even exuberance. It involves the notion of working with and through these characteristics.

Sanity is thus defined, in large measure, as an acceptance of the here and now of present conditions and a steering away from the human tendency to seek solutions that exist outside of an individual's particular existential legacy. In any discussion of the potential of online experience to exhibit the characteristics of psychopathology, this definition is a helpful yardstick to apply to such phenomena as online addiction and virtual fugue states.

Trungpa's conceptualization of sanity is reminiscent of other alternative spiritual traditions that were explored during the sixties. For example, Richard Alpert, who partnered with Timothy Leary and later became a spiritual leader in his own right, wrote a book called *Be Here Now*. The book's title nicely summarizes the healthy outlook expressed by accepting one's existence, warts and all, in the physical world at any given moment. What is interesting is to contrast this acceptance, a common theme in many great religious traditions, with the impulse to otherness that is characteristic of the flight into the virtual.

A bit of symbolic analysis, centering on an advertising slogan used by the telecommunications company Sprint, is suggested here. The popular advertisement opines that by using the company's new services, we can "Be There Now." "There" refers of course to our virtual presence in another place and time, metaphysically speaking.

This bit of innocuous Madison Avenue wordplay is clearly drawn from the title of Alpert's book. But it resonates so well that it almost demands analysis: Symbolic significance is revealing and invites comparison between the human potential movement's sense of "centeredness" and its mandate for living in the present with digital culture's constant beckoning and summoning to the realm of cyberspace, a summoning that seems to encourage—hypothetically at least—a kind of distraction from the tedium (or psychological distress or both) of the present moment.

What is the symbolic importance of the escapism associated with the flight into the virtual? Obviously, escaping by means of entertainment provided by television, movies, and books has its legitimate purposes. But escaping into cyberspace is something qualitatively different.

It has become a major selling point in digital culture's boosterism that being in cyberspace is, by its very nature, a state or condition that is somehow uniquely different from our quotidian existence in the physical world. This difference is in fact held out as part of the attraction. In this context, we should certainly invoke William Gibson's novel *Neuromancer*, a novel in which "jacking in" provides a pathway into a separate but parallel reality. In *Neuromancer,* this reality is fluid and transcendent, and being in it confers special powers.

The comparison to religious mystical experience is more than implicit, however blatantly desacralized. Similarly there are those who would have us believe this "other place" and the spiritual realm do indeed share certain numinous qualities. Could this possibly be the case? As we have seen, the virtual world is removed from the qualities of hereness and nowness that are part of Trungpa's definition of an anchoring psychological state of sanity because the virtual world offers both time and distance insensitivity. In this limited sense, the virtual truly does exist in another dimension. But the irony is that the flight to the virtual may not be spiritual in nature at all but rather a mock-spiritual experience that has the potential to move us away from a healthy existential grounding in the present and all of the social and psychological awareness that that grounding entails.

All of this notwithstanding, if we are going to invoke the notion of pathology, we should probably do so carefully. Contemporary life is rife with harsh and difficult features. Our time, tagged as it is with all the weighty historical drama of the new millennium, is indeed a rare time, marked by the chaos of emergent possibility as well as by the "gales of creative destruction" that economist Joseph Schumpeter spoke of. Familiar anchors of reality are fast disappearing, and our unwise use of technology is coming home to roost under the frightening rubric of nature gone awry.

That at this particular time in history there should arise a technology to transport us away from the chaos of the physical realm and offer an attractive parallel plane of existence, separated from the dynamism of our deepest fantasy and dream states only by the thinnest of veils, is somehow not surprising. That we should be tempted to tarry in this artificially and technologically induced state of grace, as opposed to confronting the environmental degradation and collapse that our intervention in the natural world has produced, is understandable (although not admirable).

As the physical realm continues along its entropic trajectory and the world macroproblem is exacerbated by denial, neglect, and further flight into the virtual, the real dangers that arise, of course, are seduction, addiction, and perhaps even a kind of pathological rapture. But even more problematic is the fact that as humans in an increasingly inhumane world, we seem to be all too easily distracted from the need to confront and wrestle with the harsh realities that we have, possibly with the best of intentions, created.

To opt for existence in the denial-laden pleasure dome that is cyberspace may constitute a kind of avoidance of the necessary and real Great Work, not John Barlow's facile "rewiring of the collective consciousness" but rather the much harder work of disconnecting denial. The latter requires far more courage and human spirit than navigating with a mouse or a keyboard. In addition, we must also confront our collective actions and their cascading effects in the cold light of logical and honest analysis.

Yellow Alert:

Massive System Vulnerability

Y2K, air-traffic-control failures, unexplained power outages, telephone network crashes. As impressive and admirable as our new digital technologies are, we often seem to lose sight of an important reality: They are also quite fragile in the larger scheme of things. Computers and communications can concentrate huge amounts of information, using it like an Archimedean lever to affect the physical world. But the more information that we aggregate, the more points of vulnerability we create. Aside from the Y2K problem, which seems to disproportionately symbolize this issue in the media, this is a subject that is not widely discussed—and probably for good reason. It makes us a bit uncomfortable to contemplate the wider consequences of the concentration of information.

Modern civilization is built upon the notion of aggregation. Our cities, once shining and new but now somewhat tarnished, were great nodes in a network of system resources that was the functional backbone of society itself, the complex adaptive system that is its infrastructure. In one sense, the new and invisible infrastructure that is the Internet models and mimics that same concentration of resources in the virtual realm with a certain amount of electronic grace and agility. It is a city of electrons not angels.

We can only stand back and admire these new virtual worlds from a fixed, behind-the-scenes perspective. And yet, recognizing that most technologies have involved a convergence of our resources, we begin in this age of random violence and terrorism to appreciate the virtues of true decentralization. Even though the Internet, along with other digital technologies, has the power to aggregate information, its infrastructure is more decentralized than centralized. So, in the technological pantheon, the Net plays by its own special set of rules.

These powers of aggregation and the vulnerabilities that they create will only increase as the technological diaspora continues. The miniaturization of computer circuitry and the realities of Moore's Law are driving the application of computer capability to a wide variety of uses, from embedded chips essential to the functioning of automobiles to the proposed use of computer chips as tracking devices for people. An example of the latter, proposed by David Goodtree of Forrester Research, would involve the use of computer chips to track elderly people with Alzheimer's disease or wandering farm animals. But these applications are less cause for concern than the use of networks and networking, which are far more dangerous from a social and cultural standpoint because, as Kevin Kelly assures us in *Out of Control,* we are indeed connecting everything.

As we make more connections, it is the Internet that will allow this transformation to be rationalized and optimized, to invoke a bit of Silicon Valley parlance. From one standpoint, the Internet is not so much a network as it is a common computer language that promotes interoperability between devices via the Internet Protocol, also known as IP. The interesting sidelight here is that the devices that are connected to one another do not have to be computers as we know them—they just have to have the requisite IP software. The installation of this software, a relatively simple makeover, is the price of entry (along with a few other technical adjustments) into the worldwide interconnected computer backplane that is called the Internet.

Several years ago at a trade show called Networld plus Interop, an annual pilgrimage for Internet insiders, one of the feature attractions was a toaster that had been rigged with IP capability so that it could be controlled via the Internet. This capability gives rise to a vision that provides overenthusiastic Internet aficionados with a conquer-the-world mentality. In this vision, the Internet connects not only all the computers of the world but also an incredible array of ordinary devices used throughout the course of everyday life. These devices could, in fact, be household appliances or any other kind of device that can send and receive wired or wireless transmission signals in IP format. At one Networld plus Interop show, Internet coinventor Vinton Cerf was spotted walking around with a T-shirt that said "IP on Everything."

Nicholas Negroponte's Media Lab has been heavily involved in promoting this kind of widespread application of computer technology under the rubric of Things That Think, or TTT. Very simply, the idea is to aggressively push forward the technological diaspora by installing networked intelligence into every nook and cranny of daily life. This push includes a few preposterous notions: computerized clothing and eyeglasses that can be used to read e-mail. This vision of the future hints of a world where everything is controlled and controllable via the Net or a series of subnets, where people, places, and objects are blips on the radar screen of a huge system that is an electronically managed network.

How concerned should we be about these scenarios? On the one hand, we are cautioned by digital culture that because these kinds of networks are decentralized, we should decidedly not see them as early evidence of a new form of Orwellian social control. I suspect the problem is more subtle than that and that the real question is: Are such regimes gradually disposing our society toward the acceptance of increased technocratic management? Might not such systems eventually be used in certain circumstances to collect information in a more centralized fashion?

Analyst Peter Huber, in his usual brilliant but erratic fashion, has argued that the current mode of network design, based largely on decentralization, effectively obviates concerns about the prospect of centralized Orwellian concentration and control. Yet as good an analyst as Huber generally is, he is not, in this instance, looking deep enough. From a technical standpoint, his contention is simply incorrect because when everything is networked, there is no longer a need to maintain a central repository for information. Information can easily be garnered from any number of databases or Web servers and can be traded back and forth readily. In this mode, centralization simply becomes ad hoc and, what is worse, disguised. In addition, the collection of information at any of the nodes on the network is becoming increasingly sophisticated. Companies have already found ways to, in effect, track the attention paths of those who are browsing a given Web site as they explore the tunnels of a particular information path. We should not be afraid of our technological future. But we should be aware of its potential distortions and perversions.

The notion of aggregation that is the essential feature of the computer's power in our lives has a significant downside, however: system vulnerability. When the Internet was first developed, it was built on the premise that it could survive a nuclear attack or some other type of major societal disruption. In order to do so, it had to be configured not as a centralized system but rather as a large, distributed (complex adaptive) system, so that if one part were taken out, the system would continue to function much like an amoeba would if a piece of it were destroyed. In this sense, it has a somewhat organic nature. During one of the series of earthquakes that plagued California in 1995, the Internet actually showed its stuff and performed very well compared to commercial online systems, many of which experienced significant outages.

One of the laws of network design, however, is that complexity increases as new nodes and new users are added to the system. Given the sheer size of the sprawling, interconnected global system of computers and computer-containing devices that is the Internet, some network experts worry about the ability of the Internet to "scale"—that is, to grow in both size and complexity and still be able to function at the level that its users have come to expect. Complexity and systems scalability have their limits. The Internet may be able to accommodate forty or fifty million users, but can it accommodate two or three hundred million human users and untold numbers of "Things That Think" and still function effectively?

If we believe that many corporations and other institutions—indeed many functions of society, if some predictions are to be believed—will become increasingly dependent on the Internet over time, then serious concerns about overall system vulnerability begin to surface. Network design experts like IDC's chief technologist Robert Metcalfe have predicted that the Internet will be plagued by severe outages in the coming years. In 1996, Metcalfe's predictions were borne out when major Internet Service Providers like Netcom and BBN Planet as well as commercial providers like America Online suffered significant and well-publicized outages. Y2K adds yet another dimension to this problem. Thus, in the larger picture, as our society becomes more technologically complex, its vulnerability proportionally increases.

Imagine, then, this future world consisting not only of a more complex and much larger Internet but also of a whole new layer of networked connectivity: smart devices. When we begin to talk about future worlds, our ability to continue to build these kinds of systems runs into the limitations of our conceptual ability to embrace their vastness and complexity.

From the vantage point of software development, Kevin Kelly talks about the need for complex software projects to quasi-biologically "evolve" because of the limits of human engineering. As systems become larger and more features and functionality are built in, the number of lines of programming code will begin to approach one billion or more. Kelly readily admits (and I concur) that in order to make such massive computational systems viable, we will have to develop a new level of machine intelligence that can monitor the performance of the primary system. The real question here, however, is whether this is technologically achievable or, more importantly, manageable from a human standpoint. In other words, can human technologists manage something that is by definition beyond their ken?

Is there a limit in the natural order of things to technological complexity? If so, then we must assume that the limit is determined by the ability of human intelligence to encompass or manage that complexity. At what point in pushing the envelope of science and technology do we, in our necessarily human exploratory way, get in over our heads?

Digital culture might argue that the word "natural," as I have used it here, is meaningless, since human achievement, scientific or otherwise, is constantly redefining what nature is. But if we cannot control the works of our creation, then in what sense can those creations be said to serve human interests? And, if we assume no further evolution in human intelligence, then it seems wise to respect the possibility that there are indeed, embedded in the implicate order, "natural" limitations to complexity. Perhaps Kevin Kelly's argument that we should create artificial life forms using the techniques of computational biology, turn them loose, and let them run out of control is an inadvertent warning, that is, a roundabout way of saying that science can no longer be expected to fully control its own creations.

The advent of "invisible" hyper-technologies like nanotechnology and bioengineering—technologies that can be understood only in the conceptual realm and not in any concrete way—have changed the very nature of science and technology. Because the primary and secondary effects of these technologies can no longer be successfully envisioned, moving into these realms is like ad hoc experimentation with nature itself. And what is irresponsible ad hoc experimentation if not mad science?

The risks here are obvious but not widely understood or heeded in the current environment, where technological advancement is now being fueled by business imperatives and a growth-at-any-cost mentality that has little time for reflection. These risks include the creation of ecological chain reactions, permanent and infinitely perpetuating alterations in genetic sequencing that have the potential to cause massive ecological disruption. Tampering with our genetic code should be done with the greatest circumspection, and this kind of care is a far cry from the manic and market-driven perambulations of profit-obsessed biotechnology firms.

As Einstein suggested, it may indeed be true that our technology has surpassed our humanity, especially where biotechnology is concerned. In truth, we have yet to come to terms with the consequences of the very first hyper-technology that scientists brought to fruition: nuclear fission. Native American culture is known for a special kind of tribal wisdom that looks after the concerns of its people seven generations hence. Contrast that with thinking that allows the creation of nuclear waste that will continue to be dangerous for many more than seven generations; it is a telling reversal of values.

The Internet may help us with this problem because it is first and foremost a communications technology. But if we use its universal power to automate and mechanize many aspects of life, we will if nothing else increase our technological dependency. The new prospects for hyper-technology may dazzle us, but if in the process they also blind us, how then can we guide the progress of technology with wisdom? We need to wrestle with this question: Does a strong dependency on a liberating or empowering technology represent freedom or a narrowing of choices?

The Electronic Polity

-------------- The more readily we conceive the planet as a single unit and move about it freely on missions of study or work, the more necessary it is to establish such a home base, such an intimate psychological core with visible landmarks and cherished person-alities. The world will not become a neighborhood, even if every part of it is bound by instant communication and rapid transportation, if the neighborhood itself as an idea and a social form is allowed to disappear.

—Lewis Mumford

-------------- Contemporary observers have documented and analyzed the way mass media . . . have "commoditized" the public sphere, substituting slick public relations for genuine debate and packaging both issues and candidates like other consumer products.

The political significance of [computer networking] lies in its capacity to challenge the existing political hierarchy's monopoly on powerful communications media, and perhaps thus revitalize citizen-based democracy.

—Howard Rheingold

-------------- The right to flame your congressman by E-mail is not likely to improve the quality of democracy.

—Daniel Burstein and David Kline

-------------- The political process is moving onto the Internet. Both within the United States and internationally, individuals, interest groups, and even nations are using the Internet to find each other, discuss the issues, and further their political goals.

—Charles Swett

The Complexities

of Role and Identity

in Cyberspace

Whatever you may have heard or read, the online world is not a WYSIWYG environment—what you see is not necessarily what you get. Cyberspace is teeming with hidden complexities: trapdoors, wormholes, mirages, and illusions. If you start looking for them, however, you probably will not see them—at least not right away. The complexities of the online world—especially those that involve conversations with others—do not become readily apparent until you have spent a lot of time roving its virtual corridors. Although the same could be said for just about any other activity in life, it is surprising how many otherwise well informed people have a tendency to take what transpires in the online environment at face value.

It is important to remember that cyberspace is an experiential phenomenon. Being there is what counts. Participating in discussions, making mistakes, watching others make mistakes, seeing how opinions are formed and challenged, learning the nuances of virtual social environments, watching how communications breakdowns occur, and getting a practical sense of both the advantages and the limitations of various online systems—these are all a part of learning about the way things work in the online environment. There are no ready-made analogies for human interaction in the online environment, not only because it is a new medium but also because it is a place where the ground rules are still being hammered out and the experience itself is still being shaped and invented. To a large extent, in the virtual world we are all making it up as we go along and experimenting with a new means of communication.

The hidden complexities I am alluding to come in many forms and span many situations. In cyberspace, even something as seemingly simple as identity can be fraught with complication. One of my preferred hangouts in cyberspace has been the WELL, an upscale, intellectually oriented conferencing system founded by Stewart Brand and based in California. When I first joined the WELL, I had to decide how I was going to represent myself in the community. Since at the time I was a magazine editor and a journalist who covered online issues and the Internet, the decision was even more complicated—my professional activities had a certain public dimension to them.

Everyone who joins the WELL must, with certain minor exceptions, be identified in what is called a .plan file; one cannot join anonymously. The WELL has traditionally been a UNIX-based system, and the .plan file is simply a UNIX file that lists the subscriber's full name and certain other basic information, the inclusion of most of which is optional.

When a subscriber posts comments in the various conferences, the software used by the WELL automatically attaches the subscriber's name to every posting that he or she makes. It also date- and time-stamps the posting in a numbered sequence. Posts appear with a header that includes the topic number, name of the conference, topic name, posting number, name and user ID of the subscriber who made the posting, and date, time, and length of the posting.

Topic 129 [wired]: New Republic Slams Wired!
#62 of 218: Tom Valovic (tvacorn) Sun Jan 1 '95 (19:19) 19 lines

Note that "tvacorn" is my selected user ID and automatically appears in every posting I make.

On the WELL's system, the header is fundamentally unalterable. However, whether or not my full name appears in the posting is alterable. What I or any other subscriber can do, using a simple command, is alter my full name through the use of a pseudonym. If I had chosen to do this in the case of the posting above, the header would have looked like this:

Topic 129 [wired]: New Republic Slams Wired!
#71 of 218: more icons! (tvacorn) Mon Jan 2 '95 (19:20) 8 lines

"More icons!" is a playful pseudonym that I might occasionally choose to use. However, using a pseudonym does not render my postings anonymous on the WELL system. Because the WELL, unlike many other systems, has always discouraged anonymity, information about a subscriber's true identity is always available in the .plan file. Thus, people who read my comment and found themselves wanting to know my identity could access that information simply by reading my .plan file, which is publicly available. But, interestingly, one level of complexity has already been introduced here because many subscribers will not take the time to look up the identities of those who post under a pseudonym. Thus conversations take place between the named and the nameless, and the reliable context of identity is no longer a constant.

Cyberspace is a participatory public or semipublic space that has no formal requirements for participation. Because of this, identity has significant implications with respect to the notion of electronic democracy and various other forms of private and public discourse.

Complexity of role and identity in online environments is determined by a number of factors. An individual's role in the online world may or may not, for example, be associated with a professional affiliation. For example, as a member of the WELL, I typically subscribed to several conferences including telecommunications, media, and music. When I was in the music conference, I was there strictly for recreational purposes, participating as a private citizen. But when I went into the telecommunications conference, did my relationship to the WELL as a forum for discourse change? If so, how? Because I was editor-in-chief of a magazine that covers telecommunications, should I have conducted myself differently in that conference? Did I have the same amount of freedom as someone who was attending the conference purely as a hobbyist?

Certainly, a subscriber reading my comments would have no easy way of determining whether I was speaking as an expert or as just another passerby. And how should a specific communication be interpreted? Should the critical determinant be the intention of the poster or perhaps the interpretation lent to the post by another subscriber? Thus, simply moving from one conference to another—something that people do repeatedly in a given session on the WELL—affords many possibilities for confusion of role and identity.

Similar questions arise with respect to corporate affiliation. In some cases, if subscribers participate in a work-related newsgroup and their employers pay for the online time, the subscribers might want it understood that their comments should not be construed as representing the interests or opinions of the employers.

Subscribers often deal with this situation by attaching disclaimers to an information trailer (e-mail address, company affiliation, and so forth). A typical disclaimer might read, "My views do not represent those of my company but are solely my own opinions." But what if the poster does not use a disclaimer? How should others in the conference construe his or her remarks? In the default mode, ambiguities of interpretation are always possible.

More likely than not, most people who frequent cyberspace are probably not staying up nights worrying about such niceties. As a media theorist and editor, I found myself intrigued by the inherent confusion of public and private roles that seemed to be a defining characteristic of the online experience. Besides, the stock-in-trade of editors is clarity of expression, and as I spent more time online, more opportunities seemed to emerge for confusing transactions and muddled interpretation. Furthermore, from strictly a personal standpoint, I preferred a position of clarity to the ambiguity that is always possible in online conferencing, and I worked hard to make sure that my own comments were clear and readily understandable, even though that goal was often at risk. As an editor in a very visible industry known to many participants on the WELL and other places I frequented, I also wanted to safeguard my right to speak as a private citizen.

Unfortunately, there seemed to be no surefire mechanism for accomplishing these goals. Perhaps, as digital culture is fond of pointing out, cyberspace really was the electronic frontier—a place where ambiguity and anarchy were to be savored and enjoyed with the knowledge that the rules would come soon enough. Or perhaps journalists and editors were simply more sensitive to these issues. But if the rules were inevitable, why was digital culture always talking about how there should not be any?

It is no accident that a lot of journalists frequent the WELL. Part of the system's early marketing strategy under operating managers Cliff

Figallo and John Coate was to actively encourage the participation of writers and journalists, on the assumption that they would add to the WELL's unique and upscale intellectual cachet as well as promote the system via their influence in their spheres of activity. To do this, the WELL gave journalists a limited amount of free access time so that they could test drive the system and then presumably spread the word about its erudite, irreverent, and cosmopolitan ambiance.

Although the strategy has many journalistic complications, it worked. This classic word-of-mouth marketing strategy, along with others implemented by the WELL staff, helped to raise the level of discourse on the conferencing system; and that, in turn, attracted other high-profile subscribers.

However, this practice raises a lot of questions. Might not subscribers who happen to be wandering around in a conferencing system want to know that there is a journalist lurking on every virtual street corner? Might they not have the right to know that one of their fellow conference participant might be a journalist? After all, the WELL administration did not require anyone's .plan file to include information about his or her profession. Including this kind of information was purely voluntary.

Imagine a scenario, then, where a subscriber unknowingly engages in an e-mail conversation with someone who happens to be a newspaper reporter in the subscriber's hometown. If sensitive subject matter involving an area covered by the reporter was introduced, then to say the least a lot of complications could ensue. But the WELL itself offered no guidance about such matters, nor were many guidelines likely even formulated at that stage of the game.

This scenario raises many questions. Do journalists have an obligation to identify themselves when they are roving around in cyberspace? If so, how can this be reasonably accomplished? If they have no intention of exercising their prerogatives as purveyors of public information, should they somehow make that known? If they do try to make it known, how can they ensure that conference participants will be fully aware of it? After all, surfing the Net while spewing out legalistic disclaimers at every turn of the conversation defeats the purpose and dampens the enjoyment of online participation.

These are just a few isolated examples. In general, conversation and the exchange of information in cyberspace are full of such complications, protocol complexities, and ethical dilemmas, although most participants do not seem to worry too much about them. In my own experience, many of these issues seemed to reach critical mass in the WELL's media conference, a popular hangout for editors, reporters, and journalists of all stripes. Many of my own thoughts about such matters were forged and formulated by participation in that conference, and I explored some of them in an article I wrote for *Media Studies Journal*.[1]

While working on the article, I garnered the opinions of a number of conference participants, including *Time* magazine's Philip Elmer-DeWitt and *The New Yorker*'s John Seabrook, on the ticklish subject of online media criticism. Among other things, the article explored some of the murky issues associated with role and identity. For example, is the role of journalists online primarily determined by who is paying for their time on the system? If their time is being paid for by their employers, are they then participating in an official capacity? And if so, then what does "official" mean?

Journalists might think they are speaking for themselves, but someone reading their comments might assume the opposite. For example, did Philip Elmer-DeWitt's comments in his *Time* cyberporn cover story—which aroused a storm of reaction from other journalists in the WELL's media conference—reflect his own opinions or were they to be construed as official communications from a member of the magazine's staff?

Another complicating factor is the nature of the conferences themselves. When I first joined the WELL, I consciously avoided topics that were related to my professional sphere, since I perceived my own presence on the WELL as a matter of personal, not professional, exploration. Over the years, however, conferences like telecommunications, wired, information, and virtual community, conferences that yielded interesting and useful professional content, gradually managed to insinuate themselves onto my automated conference list. But, in general, I found that the online experience created a discernible blurring effect whereby it was not always possible to say that my participation in a conference was either strictly professional or simply the dabbling of a hobbyist or a concerned private citizen. What is interesting here is that the online experience tends to break down the barriers that typically delineate the somewhat artificial partitioning of time into the neat categories of work, play, relaxation, and so forth. Or perhaps it might be a bit more accurate to say that the online environment has a way of homogenizing work and play to the point that separating the two becomes increasingly difficult.

I have used two examples here—the WELL and the field of journalism—to show how ambiguity can thrive in the online environment. However, they are just that—examples. Such ambiguities arise in many other conferencing systems, including Usenet, America Online, Prodigy, and CompuServe. They will in all cases be specific to the individuals involved. A law enforcement officer spending time online may experience one set of complications, a social worker, another. The point is that these ambiguities tend to arise at the intersection of public and private areas of interest, and the virtual world has a propensity for conflating the two in some very interesting ways.

What conclusions might we draw from the dilemmas, ambiguities, and contradictions that the online experience can engender? First of all, it seems clear that the free-form nature of the online environment tends to break down the convenient separations of function and custom that delineate the contours of our personal and professional lives. As will be discussed, this has both positive and negative dimensions. In general, however, online transactions and communication often take place in the absence of meaningful context. I refer to this phenomenon as decontextualization. For example, an online posting, however interesting or valuable, is decontextualized if its source is anonymous. As a part of the decontextualization process, form, content, and source—all essential components of any communications pattern—can be decoupled or scrambled. (And what has been scrambled can, of course, be resequenced!)

Decontextualization is a hallmark not only of the online world but also of our contemporary media. In the online environment, however, the scrambling of a signal is far more pronounced. There is a decidedly chaotic quality to the online world, which is at once its best and worst characteristic. A media theorist like Avital Ronell would likely underscore the positive aspects of decontextualization by pointing out that this "scrambling of the master codes" is necessary and useful. In Ronell's view, we can see the Net as a tool for stripping away repetitive and reinforcing societal strictures, a tool for liberating exploration of sense and self. Some of these notions are also explored by Sherry Turkle in her book *Life on the Screen*.

In this context, the Net may be seen as an instrument for social change precisely because of its power to decontextualize and scramble sign and symbol. As cultural historian William Irwin Thompson has pointed out, chaos in the old order always precedes the organization of the new order. The breakdown of old forms and modalities to make way for new ones is an essential part of a society's progression. In the final analysis, however, the real question is whether the Net can provide a means of reintegration as well, a means whereby new forms and structures can emerge from the chaos of transition and allow the Net to evolve into a new and radically different vehicle for public and private communication.

Spin Doctors

Invade Net—

Film at Eleven

One of the more interesting hidden complexities of cyberspace has to do with how opinions are formulated, shaped, and presented. In this context, it is important to examine how bias can exist in the virtual world, just as it can in the realm of print and television media. It is also critical to identify how the expression of free speech can be controlled, especially since there are those who feel that the Net has become a place where free speech is in its unfettered form. Being able to exercise free speech is one thing. Seeing how that exercise is subtly shaped, modified, and constrained by various media venues is another.

Is the online environment the ideal venue for free speech, as some say that it is? Here again, there are hidden complexities. We find, not surprisingly, that in the virtual world, systems tend to favor system users: The more a subscriber burrows into the culture of an online system, the more adept he or she will become at using it and negotiating the complexities of the virtual terrain. A good way to understand this is to get involved in a few controversial debates on a well-designed bulletin board system. Experience will show that the advantage does not always accrue to the person whose arguments are most effectively presented. An important factor that is frequently hidden from the casual user is the amount of time that one has to spend defending one's position.

Again, I will use an example from the WELL. As in other online venues, posts on this system gradually accrue in sequence over the course of time. Discussions can go on for days, weeks, or even years because comments are posted asynchronously. In other words, a new comment in a discussion might be appended to a sequence of existing comments at three in the afternoon, two more might appear at around half past three, and four more might show up between four and five in the evening. In controversial discussions, participants who spend the most time online watching the discussion unfold have the advantage. In a heated argument, several hundred comments or more can appear in the course of an hour. If you are involved in such a discussion but do not happen to be around at the time that it takes place, the train will leave without you. If you are being attacked because of a previous statement and are not around to defend yourself, you may come across very poorly if your opponents happen to have the luxury of being present during the relevant time span. In general, the nature of the online experience favors those who spend the most time in cyberspace. Online habitués are the most likely to enjoy the advantage of the last word in arguments, to be able to respond to criticism directed at their posts, and, most importantly, to shape the flow of discussion. And here is the most interesting point in all of this: On a system like the WELL, where service is metered on an hourly basis, those who can afford to spend longer amounts of time (and more money) on the system are in a position to enjoy the benefits of online free speech more than others. Thus, there is a decidedly arbitrary quality to the virtual world that is not always appreciated by its more casual users.

Given this context, how democratic, then, is electronic democracy? Once again, economics has significant implications with respect to electronic democracy and how public discourse evolves in the virtual world. If we can cynically state with respect to print media that freedom of the press belongs to he or she who owns a press, then we can with equal cynicism state that freedom of the Net belongs to those who can afford it. As long as online services cost money, freedom of expression will be granted in unequal proportions to those who are most able to pay for it. (Fortunately, most online services now offer unlimited access for a fixed price; but this model may not remain in place indefinitely.)

Anyone who has spent a considerable amount of time online can testify to the fact that the online experience is self-reinforcing. If you spend enough time in a given conference, you can outlast opponents simply by wearing them down with your comments. Unfortunately, this technique

----

can give a single well-heeled individual or group of individuals a surprising amount of power and influence over the shaping of dialogue.

This phenomenon is easy to observe on the WELL and other online venues once you are sensitized to it. In some cases, individuals who get the last word are those whose employers subsidize their time on the system or those who for other reasons have more time to invest in online discussions than others.

I personally elected not to partake in a number of heated arguments on the WELL about issues that I cared passionately about for one simple reason: I had a life and did not want to be tethered to the computer for hours at a time simply to reasonably defend a controversial statement. But the reality is this: People who do not take the time to monitor a discussion in which they are primary participants run the risk of being personally attacked or having their arguments discredited.

In general, there is a bothersome lack of accountability in the online world. The sniper can all too easily gain the rhetorical advantage. Furthermore, it is altogether too easy for a single persistent individual to derail the course of a discussion by throwing a few conversational grenades— items that are bound to take the discussion into undesirable territory. I have seen many fascinating, worthwhile, and unusual discussions get sidetracked in this fashion. This phenomenon can be one of the more disappointing aspects of the online experience.

Once topics do veer away from the core issues of a discussion (an occurrence known as "topic drift"), bringing them back is almost impossible. Imagine that a subscriber has started a discussion of, say, President Clinton's place in history. Once the topic has been started, there is absolutely no guarantee that it will stay within the confines of its intended borders. The topic will, in fact, be defined by the sum total of the posts of each participant. If the discussion drifts far enough and the new material is more interesting than the original, then it may become virtually impossible to address an earlier point.

Once topic drift has occurred, a subscriber or the topic originator can force the issue and protest that the thread has been hijacked or derailed; but the fact is that in the delicate ecology of online discussion, there is really no way to go back. It is worth noting here that topic drift is not always a negative phenomenon; some of the more interesting discussions on the WELL and other systems have serendipitously resulted from drift. However, topic drift may be the reason that much online discussion remains fixed at a relatively superficial level.

The nature of online dynamics is too often taken for granted and viewed as a simple sequence of linear comments. The reality is that what happens online is part of a complex, decontextualized process that can be difficult to deconstruct at any given point. This complexity raises a lot of interesting questions, including questions involving the medium's potential for the manipulation, rather than just the expression, of opinion.

In theory, subscribers with the most-favorable access—which we can define as being a function of both technological ease and freedom from economic pressure—can bring more influence to bear online than others. Having both access and the time to invest in getting to know the inner workings of cyberspace and the quirks of particular online systems is a distinct advantage with respect to the ability to "freely" express opinions. On the WELL, this advantage is commonly known and appreciated by the more cybersavvy subscribers. In fact, Howard Rheingold and others have referred to the WELL as a "posthocracy," meaning simply that those who post most frequently do indeed have the most power and influence online.

The architecture of the WELL as an example of the New Media is highly relevant here. The WELL is an aggregation of conferences, and each conference has one or more hosts. What is most germane in terms of free speech, however, is the power that is conferred to hosts and how that power affects online interactions and the rights of others within the system to express their opinions.

While hosts are generally not paid members of the WELL staff, they are given compensatory time and significant system privileges, both of which provide considerable freedom of movement and a kind of strategic awareness of the conferencing system's rather complex information landscape. Among other things, hosts have the power to guide discussions; "freeze" topics, that is, censor or stop discussions that are deemed inappropriate to the WELL community; forbid topics altogether; and even, in extreme cases, ban problem subscribers from their conferences.

In addition, hosts also have significant archival powers that allow them to eliminate outdated conference topics. Material is removed from the current topic list at the host's discretion for the most part, and the decision as to what constitutes outdated seems to be purely subjective. Politically incorrect discussions (depending of course on the politics involved) can easily be removed from the roster of topics, and users are rarely in sufficient agreement to prevent such actions, even when those actions are unpopular.

I do not mean to suggest that WELL hosts abuse their power—far from it. For the most part, hosts seem keenly aware of their privileges and do their best to adjudicate conflicts and use their influence within reasonable limits. (In a way, it is the structure of the system that reestablishes hierarchical control.) I do wish, however, to stress that their power over content is considerable, even though it is neatly hidden, albeit in plain sight, from the user community.

In making this point, I hope to further dispel the notion that all online venues are strictly egalitarian playing fields for free speech and democratic discourse. They are not, and they are not likely to get any more democratic in the future: Even now we watch the ascendancy of gated communities in cyberspace, be they private conferences on the WELL or partitions of the electronic agora.

In some respects, power over content in the online dimension is analogous to that which broadcasters have with respect to television programming. There is, however, a distinction that needs to be drawn carefully. Even though hosts do not have significant direct influence over the content of ongoing topics (with certain important exceptions as stated earlier), when it comes to saving material already discussed and eliminating older topics, their power is indeed significant and almost arbitrary.

Even though many hosts courteously let conference participants know in advance which topics are slated for elimination, the process is flawed in other ways. First of all, users who are off the system for an extended period of time might easily miss such an announcement. Second, the process tends to put subscribers in a supplicant position—they need to specifically request that old material be kept and, if not asked for their opinion, may hesitate to give it. Eliminating older topics when they are controversial can amount to a kind of censorship after the fact, although it is difficult to prove that such an event or act has occurred.

The reason I have taken such pains to delve into these complexities is to challenge what has rapidly become the conventional wisdom about the virtual world: that it is a means of creating equality, that it is egalitarian in nature, a haven for free speech (indeed digital culture would have it as one of the last remaining bastions of free speech). At a minimum, I hope to have demonstrated that these generalized notions do have some degree of truth but that when we look deeper into the nature of electronic transactions, these notions can no longer retain their validity as simplified truisms.

On the surface, the fact that anyone can publish on the Web or post in a bulletin board appears to offer individuals an unprecedented level of access to the emerging electronic polity. But to assume that this freedom is unmitigated by external forces, or will remain so in the future, is rather naive. Digital culture is fond of criticizing the traditional media as being controlled by the narrow interests of editors, owners, and publishers, who filter information before it is presented to readers and subscribers. The title of an article that appeared in *Wired* written by Jon Katz, "Online or Not: Newspapers Suck," describes this perspective nicely. The fact is, however, that power similar to that exercised by the traditional media can and does exist in cyberspace, especially with respect to online conferencing systems. It resides in the hands of system administrators, conference hosts, bulletin board sysops, newsgroup moderators, and others who often have little accountability for the decisions over the disposition of content that they invariably make. In addition, the invisible wires and ropes that control cyberspace content are often not apparent to the casual user. It is only through painstaking deconstruction that they are revealed.

In addition to the fact that they are compensated for their work online, WELL hosts also have a large array of UNIX commands available to them in the course of their daily ministrations over the content of the conference. The capability afforded by these commands can provide them with, among other things, discretionary information about how users spend their time on the system, about users' online habits, and even about personal lives in ways that many WELL users are probably not even aware of. (In fairness, however, many of these commands are available to regular users also, although it takes time and technical savvy to become proficient in using them.) In the WELL's meritocracy, technical knowledge is rewarded with an enhanced ability to navigate the system. Although WELL hosts seem in general to be conscientious about their responsibilities, the potential is always there for abuse of these powers.

Like it or not, there are indeed invisible hierarchies and power structures in cyberspace. It could be argued that in a way these structures pose more concern than their print media counterparts precisely because they are not as immediately visible. In a community-oriented system like the WELL, these hierarchies tend to take a different form than what might be seen on online systems like Prodigy or America Online.

While some of the principles that I have described regarding the WELL are not necessarily universally applicable to other online systems or to the Internet itself, many of them are. We are reluctantly forced to conclude then that, in general, cyberspace is not in all cases and circumstances the egalitarian platform for free speech that the conventional wisdom would have us believe. Its much vaunted freedom is bestowed in proportion to the technological ability of its users and their ability to muster the time and financial resources to play the game.

If some users do indeed have more power than others, then some even more interesting and tantalizing scenarios emerge. For example, could a single individual dominate a Usenet group, a conference on the WELL, or even the WELL itself? "Dominate" is, of course, a loaded word and requires clarification: I am using it here in the broad sense of "to unduly influence." But the thought experiment needs to be performed.

What would happen if a single articulate and entertaining person on the WELL began posting in profligate fashion (that is, began to try to influence content by the sheer number of postings)? Group pressure might be brought to bear on the poster to moderate his or her involvement. Sometimes this tactic will discourage individuals from overinvolvement, but at other times it has the opposite effect. In any case, even this rather crass and obvious tactic can give a single poster significant influence on the final direction of a topic. There are, however, a number of more subtle ways to influence opinion online, if that happens to be one's objective. As we have seen, the key to reaping the rewards of online participation is simply to increase one's involvement. To be truly effective and influential at shaping content and opinions, a subscriber needs to invest both time and energy online. If a subscriber has the time to follow a topic, evaluate the twists and turns of the discussion, and then compose responses that have an impact on that discussion, then the subscriber can be effective in influencing opinion.

All of this, in turn, raises an interesting prospect, one that seems to be little discussed in the annals of cyberspace: the notion of online spin doctoring or opinion shaping in support of specific objectives, either by individuals with vested interests or special interest groups. Could spin doctoring happen on the Net? The chances are that if it has not already, it most certainly will. When it does, what will it do to the credibility of cyberspace as an information source, and what effect will it have upon the quality of information?

Let me ratchet up the paranoia level by offering the following example. Imagine that some special interest group—say a political splinter group, a lobbying organization, a foreign government, a racist hate group—wishes to influence online opinion. Here is how it might work. An individual representing the group stealthily signs on to an online system. If it is an anonymous system, so much the better. This individual—acting on behalf of the group—would then be free to visit any number of conferences where relevant discussions take place and engage in the time-honored practice of generating propaganda. With sufficient time, energy, and skill, he or she could indeed influence opinions. If that same special interest group were to hire several people to engage in this type of activity full-time, the impact on a relatively small system like the WELL would be considerable, and on a larger system or systems, significant.

Under the currently loose and open guidelines that prevail in cyberspace, there is little to prevent this from happening, and there are few viable mechanisms for ensuring that this kind of dubious opportunity is not taken advantage of. The ease with which this kind of maneuver could hypothetically be accomplished is a bothersome detriment to the "purity" of cyberspace as a means of civic discourse. At a time when information warfare is increasingly utilized in both the corporate and political spheres, this type of propagandizing has some serious implications for the notion of fairness of presentation in an electronic democracy. If and when this begins to happen on a large enough scale, a huge "tilt" sign will light up in cyberspace, and the environment will have been changed forever. In cyberspace, unlike the traditional media, there is at present no good way of ascertaining the veracity of sources.

I have offered a hypothetical example, but there are a few actual ones available as well. In 1995, the engineering staff of a large California-based network-equipment vendor flooded a technical Internet newsgroup. The staff's mission was to counteract negative publicity that had appeared in a trade magazine: One of the company's networking products had received unfavorable ratings in a performance test. Did the company put its employees up to this task? And if so, what are the larger implications of such an action?

In another case, the WELL was embroiled in a situation involving a lobbying organization, the Electronic Frontier Foundation (EFF), founded by Mitchell Kapor and John Barlow. The EFF was smart enough to recognize that the best place to influence online policy is online. Accordingly, the EFF's online counsel, Mike Godwin, was a very visible presence on the WELL. Godwin spent a lot of time in places like the media conference representing EFF policy positions. What is unfortunate is that it was never quite clear whether the opinions he presented were his own or the EFF's. Although many of the EFF's positions on civil liberties in cyberspace are worthwhile (such as its stance against the Communications Decency Act), the worthiness of its cause is immaterial to the question of online propriety that needs to be asked here.

This problem, like others we have seen, goes back to the issue of role and identity in cyberspace. Godwin's online activity (as well as his salary) was subsidized by the EFF—an organization with specific objectives and a vested interest in shaping public opinion and ultimately influencing Washington's Internet policy making. Although the ambiguity raised by Godwin's postings is highly problematic, his involvement as an EFF representative raised few eyebrows on the WELL simply because many subscribers were highly sympathetic to EFF positions; for them, the end seemed to justify the means. However, if the Michigan Militia or the CIA were to do the same thing, there would undoubtedly be a huge outcry.

The point is that by subsidizing Godwin's online time, the EFF conferred power to Godwin over and above that of users participating on their own nickel. The free speech advantage, thus, goes to Godwin and his organization in certain circumstances, and this fact plays havoc with the notion of equal access. If a casual user had tried to engage Godwin in an argument in the media conference, he or she would probably have lost because Godwin enjoyed the benefit of having a lot more time to construct a defense and argue his points—unless of course that particular casual user had been similarly subsidized and able to spend a similar amount of lavish online time without worrying about the financial consequences.

Thus, one of the biggest flaws in the notion of cyberspace as an electronic democracy is its strange obscurantism: A subscriber never really knows who has made a posting and what his or her motivation might be. The problem raises the possibility that government organizations either in the United States or in another country—the Internet is, after all, a global system—might take advantage of this flaw and be tempted to use the Net as a propaganda device. In a Department of Defense report entitled *Strategic Assessment: The Internet,* this possibility is in fact explored:

It would be possible to employ the Internet as an additional medium for Psychological Operations (Psyops) campaigns. E-mail conveying the U.S. perspective on issues and events could be efficiently and rapidly disseminated to a very wide audience. The United States might be able to employ the Internet offensively to help achieve unconventional warfare objectives. Information could be transmitted over the Internet to sympathetic groups operating in areas of concern that allows them to conduct operations themselves that we might otherwise have to send our own special forces to accomplish. Although such undertakings have their own kinds of risks, they would have the benefit of reducing the physical risks to our special forces personnel, and limiting the direct involvement of the United States since the actions we desire would be carried out by indigenous groups.[2]

If you consider that the number of WELL participants who post their remarks in a conference like politics is actually quite small, one of two foreign agents dedicated full-time to an advocacy or disinformation campaign could, in effect, completely destroy the validity of a discussion without being detected by any other conference participants. The same, of course, is true of other online systems. Caveat emptor.

The Strange
Obscurantism of the
Virtual World

As one who has made a living as a wordsmith and has cultivated an appreciation for precise language, I find communicating in the online world to be a highly imperfect exercise. Online communication is communication once removed; it takes place through a glass darkly. The gestures and articulations made in cyberspace are rough hewn by the constraints of the medium; they are almost analogous to the sound bites used in the broadcasting environment. Comments and postings in cyberspace are easily misinterpreted, and efforts at clarification are often more trouble than they are worth.

This built-in imprecision in the communications process can be overcome, but only by hard work, determination, and a willingness to use language in a way that most of us do not really wish to bother with. If a posting causes confusion and a particular issue or communications snag cannot be summed up neatly and succinctly, discussants rarely attempt to keep working on the misunderstanding until they get it right. The impulse is always to move on and scoop up the next batch of sound bites.

I have found this to be the case even in communications between writers, in which one would expect clarity of expression to be paramount. In other words, the phenomenon seems to have little to do with the language skills of participants. As a communications tool, the medium itself is simply not a sensitive instrument.

This is not to say that important communication does not take place in the virtual realm. When juxtaposed against the immediacy of other communications media, such as the telephone, the asynchronous nature of online systems does offer a distinct advantage: It makes it possible to leisurely compose a response offline and then upload it to a discussion. When this kind of care is taken and participants use their writing skills to the fullest, useful and interesting discussions can and do occur.

The highest quality discussion occurs, however, when posts are accompanied by contextualization, as for example, when the identities of the discussants are reasonably well known. Imagine a television debate on a political issue in which the participants are unidentified and they all take different and extreme positions on the issue. Viewers would not know what organizations the individuals were affiliated with and thus would have no way to account for bias. They would hear positions being put forth and defended, but the decontextualization involved would render the debate confusing and one-dimensional, flat. It is really a figure-ground problem: The viewer is unable to compare the figure, the participants' comments, against the ground, their organizational bias.

Discussions in the online world are similarly flawed. We can hear strong, clear, and interesting voices emerge from the anarchic pluralism of the Net, but those voices are still disembodied. With some hard work, research, and a good "ear," one can unearth bias and background to an extent great enough to provide more depth and context to a discussion. But, of course, most participants have neither the time for nor the interest in this kind of recontextualizing. This means that for the most part the only way to deal with online discussions is to take them at face value, thus surrendering quality of information.

This picture becomes even more complicated when the role of offline relationships is considered. On any given online system, some participants will know one another personally and others will not. In a regional system with a strong national following, like the WELL, there are two distinct classes of participants: Those who know each other offline and in real life (IRL) and those who do not. It is impossible for a participant to know how this complicated nexus of relationships might affect the content of various conferences and where the hidden agendas might lie.

One specific example of an offline relationship is marriage. There are a number of married couples on the WELL. If you spend enough time on the system and pay close attention to clues and nuances offered, you might figure out who those couples are. Having done this, you would then naturally tend to factor this knowledge into your assessment of their comments, especially if there happened to be strong agreement between the two parties. Let us return to the television debate analogy: Imagine that two of the debaters were married and the audience did not know it. Once again, bias and context emerge as critically important in making good judgments about the quality of information presented. When seen in this light, an online system like the WELL would seem to favor the following "power users":

- hosts on the system
- influential individuals in the nonvirtual world
- individuals who personally knew others on the system and thereby gained the benefit of context
- individuals who spend a lot of time on the system.

To the ordinary subscriber on a system like the WELL, many of the comments and discussions that make sense to participants in one or more of these categories are marked by a kind of obscurantism. Subscribers who are outside of the inner information circle are, of course, at a distinct disadvantage.

It is difficult if not impossible for casual users to discern all of the connections between participants, which means that the system increasingly gains in richness, nuance, and context for the power user but remains one-dimensional for the casual user. Hierarchies can and do exist in cyberspace, and, as in real life, power is not evenly distributed. Unfortunately, many users are unaware of these nuances because they have made the mistake of taking the online world at face value.

The notion of complex and sometimes invisible hierarchies in cyberspace is fascinating but difficult to deconstruct. Group interaction and behavior in cyberspace have not been extensively studied simply because of the newness of the phenomenon (not to mention the difficulty of conducting field studies in virtual space). But just as individuals can shape opinion and steer discourse in cyberspace, so too can groups, especially when conference participants are unaware that a group even exists.

Interestingly, depending on the structure of an online system, groups can form and associate outside of the commons in the virtual equivalent of private clubs and can then return to the commons with certain assumptions and understandings in place. There is nothing intrinsically negative or insidious about such activity, since it reflects a natural pattern of social dynamics in the nonvirtual world.

On the WELL, for example, an interesting phenomenon began taking shape as the system grew in popularity and became more like a bustling virtual city than a small community. As the number of anonymous new faces and voices grew, many of the original subscribers on the system—those who had helped to define its culture—became disenchanted with these changes. In a kind of virtual version of there goes the neighborhood, there was an episodic but noticeable retreat from the commons and into less visible, more private conferences. In the mazeways of cyberspace—and there are many—groups and individuals can very easily retreat. The very complexity of cyberspace allows and encourages this kind of burrowing into deeper labyrinths and levels of obscurantism. There are wheels within wheels in cyberspace. And, as in the nonvirtual world, less participation in the virtual commons diminishes the vitality of civic dialogue.

It may be useful to look at this phenomenon in light of the contrast between traditional print media and the New Media. In the traditional media, figure and ground are firmly established. Opinions are rendered as such, and they have a face, a name, a context, and perhaps even a history. (We may not agree with the viewpoints presented by the traditional media, and, as Jim Fallows points out in his book *Breaking the News,* the quality of reporting has measurably diminished in recent years. However, these facts should be considered in this context as separate issues.) There is a consistency of format in the traditional media that gives them depth and scope and allows the reader to contextualize the information presented.

In cyberspace, however, relativism rules. There is little to anchor the information commons, and anonymity, inconsistent participation, and dubious motives can hopelessly complicate the contours of any given discussion. All of this suggests that we should not, as digital culture is so eager to do, abandon our traditional forms of information dissemination and lunge headlong into the admittedly exciting environment that is the electronic polity, at least until it has proved that it can deliver comparable levels of quality of information.

While the chaotic nature of discussion in cyberspace, its so-called anarchic pluralism, is one of its best features, it is also one of the worst. As an informal means of discussion, cyberspace has many virtues. But when it is held up as an alternative to traditional mainstays of society, such as the voting process or the discussions that take place on the op-ed pages of major newspapers, then it simply cannot compete on the same ground, let alone win, at least until further regimes and structures are developed to enhance information quality.

Another myth of digital culture is that information presented in cyberspace is somehow more empirical than information filtered through traditional media hierarchies. In fact, the opposite is often the case. While there is no question that important and worthwhile discussions can and do take place, opinions in cyberspace are simply that, opinions. It costs very little, in the personal sense, to render an opinion, and there is no accountability structure in place that can allow comparison between the gravitas of more formal public discourse to a free-floating, shoot-from-the-hip forum like Usenet.

The conventional wisdom of digital culture holds that discussion conferences are somehow analogous to the peer-reviewed discussion that takes place within the scientific community. The thinking is that in cyberspace, ideas are tested by being subjected to counteropinions in the heat of mixed discourse; whereas in the traditional media a single individual—a reporter or an editor for example—has the final say.

While this is a tempting notion, the fact is that the problems of decontextualization and randomness that affect Net discussion in general still apply. The gratuitous Sturm and Drang of cyberspace give-and-take is no substitute for the assiduous research that exists in the domain of science or traditional journalism, both of which at least recognize the goal of objectivity even if contemporary thinking leans toward the notion that it is unattainable.

The notion of accountability is another hidden complexity that is generally unrecognized. In the fashionable obscurantism of cyberspace, there is no consistent mechanism by which an individual can be held accountable for his or her statements and opinions, other than the cultures in various communities of interest.

While some would argue that this lack of accountability allows discussion that is more honest and more open, the negatives seem to outweigh the positives. A conference participant can inject a highly controversial comment into a discussion and then refrain from commenting for the remainder of the topic. This lack of accountability often leads to negativism and verbal sniping instead of reasoned discourse. There is safety and security behind the computer screen. Hit-and-run comments can be made with impunity, and follow-up questions need never be addressed.

The communing of disembodied spirits that seems to be a dominant metaphor for online interaction offers some interesting possibilities. For example, a number of studies have compared participant interaction in online meetings with that in face-to-face meetings. These studies, such as the one conducted by Sproull and Kiesler at MIT, found that the two environments tended to yield very different results and responses. In the online environment, rank and hierarchy are stripped away, and what is left is the context of pure ideas, vying for acceptance in what is theoretically a more egalitarian setting. In cyberspace, no one knows whether you are a dog, the CEO of General Motors, a member of a minority group, or an automated computer script.

Anonymity does offer positive dimensions. It allows online participants to express themselves freely in a way that might not otherwise be possible. It offers the cybernaut a chance to step into the realm of free expression and instills the enjoyment of what might be called pure mind play. In MUDs and MUSEs, identities and roles are tried on for size, and different perspectives can be worn like a new hat. For some people, anonymity seems to free up their expressive capabilities in unique ways.

Sociologically speaking, cyberspace anonymity has certain parallels with the rites of passage associated with urban migration, whereby individuals are freed from the constrictive boundaries of a more community-based sense of self. *Virtual City,* the name chosen for *Newsweek*'s now defunct magazine about online culture, hit a resonant chord in this respect. Some theorists argue that, like the urban environment, the online world provides an opportunity for increased sophistication, for the sharing of ideas with like-minded individuals, and the realignment of one's sense of identity in a free-form environment.

As a construct, cyberspace offers the total freedom that only a construct can offer. Because it is virtual and fluid, experience in cyberspace bestows the feeling that one has access to the levers that control and define consensus reality and that participants can be anyone or do anything they wish.

The sociological icons for our age are the bumper sticker and the T-shirt, and they are a particularly effective distribution system for scattering memes across the landscape of popular culture. Those icons also symbolize the new information age itself, in which we now proudly display our personal information as a badge of honor, albeit in the trappings of pop iconography.

In postmodern terms, the removal of the overarching metanarrative that has traditionally framed and structured our sense of society, has encouraged a new and spontaneous wave of personal expression. With the advent of the Internet, the lid has come off of a Pandora's box of free expression. We are undertaking an interesting (but probably not unique) experiment, one that has an unknown but undoubtedly fascinating outcome.

As we struggle in the transition from one societal mode to another, which has yet to be defined, decontextualization is a cultural change agent. Before a new societal framework can be established and recognized as such, old forms and structures have to dissolve and fall away. In the epistemological crisis that is the postmodern and postindustrial condition, the tendency is to question not only what we know but, more importantly, how we know it. In this quest, society's very assumptions are up for grabs and ripe for redefinition. The Internet, as a distinctly postmodern phenomenon, reflects and contributes to these trends with the unparalleled richness of its pluralism, infinite variegation, and prismlike refraction of truth and viewpoint.

Virtual Schmoozing:

The Ever Popular

Cocktail Party Effect

Can you trust the information offered by others in online discourse? If, for the sake of argument we assume, incorrectly, that the online world is a WYSIWYG environment, then we must also assume that all participants approach cyberspace with equal sincerity of intention and that their motives are relatively pure. By "relatively pure," I simply mean that individuals posting comments to a conference are representing themselves, that they are stating their views on a subject honestly and openly with the intent to affect but not manipulate opinion, that they are not deliberately posturing or engaging in role playing (which is quite common in cyberspace), and that they are not sporting a hidden agenda that overrides the scope and nature of the discussion. Many online participants easily fail this test. In fact, many who participate anonymously in conferencing systems do so with the precise intention of trying on other personae and experimenting with their sense of self.

The conventional wisdom is that we should not expect social interactions in the online world to differ from those in real life. Thus, because a certain amount of deception, social posturing, and dissembling takes place in real life, we should not be surprised to discover the same takes place in the virtual world. Likewise, if there are complex motivations and hidden agendas in real life, then those same characteristics can be expected in online interaction.

As is often the case, the conventional wisdom is pointing in the wrong direction. The hidden complexities of the online environment suggest otherwise. The online world has generally been far too idealized as a place where some ill-defined notion of pure communications takes place. However, if you give people their own electronic printing press and

window into a new virtual world, they will each approach this new potentiality differently. Some will see this tool as a means of pure information-
gathering. Others will view it as a means of trying to reestablish a sense
of community in an increasingly harsh and disjointed cultural environment. Still others will focus on it as a place in which to float their political views across the wide audience or as a pleasant amusement or diversion,
a better way to spend their free time than watching television, with its
fifty-two channels and nothing on.

If all these stances toward cyberspace can peacefully coexist, then the
same can be said with respect to particular online venues, even specific conferences. The WELL's media conference, for example, meets different needs
for different people. Some participants see it as a place for serious journalistic
feedback. Others view it as little more than a virtual water cooler or the
corner bar.

Because there are so many possible approaches to a single conference,
these Internet "meetings" are very different from professional meetings in
real life. In the nonvirtual world, an individual might attend a special
conference for enhancing career skills but would attend a different venue
for professional socializing (such as the aforementioned corner bar). One
venue would not necessarily fulfill both functions, as is often the case in
the virtual world.

In cyberspace, people do create their own realties, but they also
bring their own special requirements to situations where they might be
inappropriately applied. Or they might simply have a radically different
view of the nature of the conference than other participants do, in which
case the nature of all communications and their interpretation would be
affected. Because of this blurring of form and function, the way that
intentionality and motive affect the quality of information remains a
critical issue.

I use the example of the WELL's media conference because I am
familiar with its complications. A lot of interesting discussions take place
there, and some accomplished writers and journalists participate. But I
have found myself alternatively delighted and frustrated by some of the
discussions.

The central question remains: Can the comments being made in the conference be taken at face value? Is this just a group of writers, editors, and journalists sitting around at a bar after hours and letting their hair down? Or is it a cocktail party at a chic urban venue where professionals are on the make and on their best behavior, trying to rack up points with the power players and possible future employers or business contacts in the room? And, if the latter, should we describe such discussions as free and unfettered or as calculated and controlled? Or could such discussions be crafted to look free and unfettered but actually be calculated and controlled?

The answer, of course, is that it depends. Among other things, it depends on the attitude that each participant brings to the conference. And probably to a certain extent all of the above pertain at various times and under various circumstances. But again, with so many possibilities, how can participants reasonably correct for hidden agendas? They cannot, and therefore they must lower their expectations and take comments at face value. But in falling back on this relatively superficial approach to the information presented and tacitly accepting the necessary decontextualization involved, the participant must also accept an unfortunate diminishment and loss of meaning.

There are many stories on the WELL about writers landing jobs via their participation there, and the media conference certainly can at times have the air of a career development seminar. In this conference, you can find a mix of accomplished professionals as well as clueless newbies and journalistic wannabes. There are freelance writers hoping to be hired by editors in the conference—writers who might be tempted to tailor their remarks accordingly. There are established authors who want to avoid appearing foolish or wrongheaded in the face of a well-respected audience of professionals.

In this mix, however, there are also individuals of high integrity who render their opinions in unvarnished, Menckenian form, with no regard for anyone else's opinion of them. These individuals are the true heroes of free speech in cyberspace, although it may not always be easy to spot them in the crowd. But with this impossible mix of agendas—

what I call the "cocktail party effect"—how does a participant, or even a lurker, adequately and effectively sort out the differences? How does a user identify motive, intention, and in the worst case scenario, obvious posturing and self-aggrandizement? Or shall we simply say that it is all part of the Internet's putative but elusive charm and the strange obscurantism of cyberspace.

A Postmodern Dilemma:

Are All Ideas

Created Equal?

There is no question that we live in, as the Chinese might say, interesting times. Many years ago, an astute social observer predicted that the United States and the Soviet Union would exchange positions in social outlook and culture. I thought the idea fairly preposterous at the time, but now I am not so sure. In an op-ed piece that appeared in *Newsweek* after the 1996 Republican National Convention, network news anchor Tom Brokaw compared the heavily media-controlled event to a Communist party gathering.

In the epistemological soup of our postmodern era, we certainly see the notion of political correctness surfacing in the oddest places. Social critic and commentator James Burke sees the movement toward political correctness as a dangerous one in that it conflates equality with egalitarianism. This new sense of cultural homogenization, which tends to gloss over the deeper sources of the problem, seems to send the message that it is no longer appropriate to recognize or speak about our cultural differences. Some aspects of the movement, in fact, seem downright Orwellian.

To what extent is the online world a part of this "managed reality"? The anarchic pluralism of the Net and the associated notions of electronic democracy have a bothersome undercurrent. Our nation has long espoused the idea that all men and women are created equal. But as we continue to fail as a society to bring about true racial integration and equal opportunity, it seems that we move ever closer toward a media environment that must hypocritically pretend the opposite. The danger is that in this new environment, and in mock deference to the anarchic pluralism of the Net, this posturing tends to reinforce the notion that all ideas are created equal.

The notion that everyone's opinion is valuable springs from a democratic impulse. But opinions are one thing, and ideas—the conceptual anchors that constitute the broad canon of our civilizational values—are another. That online conferencing, talk radio, and electronic town hall formats offer the average citizen a unique voice in social and political affairs is unquestionably a positive trend. In general, the process of democracy should involve and engage our deepest sense of participation. But as the experience with various referendum initiatives in California has shown, giving the electorate an opportunity to render an opinion should be carefully distinguished from giving the electorate a somewhat arbitrary power to make decisions about complex problems on the basis of insufficient information. The mainstream media continually dumb down this participatory form of democracy by asking people to render opinions about complex issues in the political realm.

In the new epistemology of postmodern culture, the penchant toward free expression is healthy. The fact that all groups within society are being given a voice and the means to express their views more openly is obviously also desirable. However, the related but mistakenly inferred notion that all ideas are of equal value is its associated downside. It is this notion, taking a free ride on the new forms of expression made possible by various digital mythologies, that appears to represent a subtle but discernible leap into the difficult terrain of unfettered relativism.

As we have seen, the anarchic pluralism of the virtual world serves to reinforce the notion that in a democratic society everyone's opinion is important. But it also seems to subtly condition us to accept the specious corollary that all ideas are of equal value by suggesting that all contributions should be considered with equal attention.

In a given online venue, everyone who participates may render an opinion. A common format for these venues is a sequenced and linear one that structurally seems to ensure fairness of representation: In due course, everyone can step up to the microphone and chime in with a comment.

In theory, this mechanism could be used to create a system of feedback in which the genuine feelings and opinions of citizens are channeled back into the political system. By contrast, the traditional mechanisms of politics have, like the mainstream media, tended to operate in the broadcast mode. The voting public has rarely had an opportunity to provide substantive input on issues, as opposed to simply voting to be represented by an elected official who then makes day-to-day decisions and deliberations on the public's behalf.

Unfortunately, the proper balance between some putatively increased degree of citizen involvement and the necessary act of allowing the political process to do its own work has yet to be established. Furthermore, there is every reason to believe that this balance will not be achieved. The more likely scenario is that the polling process will increasingly stand in as a surrogate for true democratic participation, which will in turn be diminished by these very same forces.

In theory, the Net is an egalitarian platform for free speech available to the populace at large. As we have seen, however, it is free only in the sense that it is freely available to anyone who can pay for it. In practice, there are many real-world complications that make electronic democracy—its logical extension—an elusive ideal.

Depending on the conversational format of a given online system, comments and opinions tend to be mixed together in an undifferentiated stew in which the sublime and the ridiculous coexist on the same surreal, discordant, and uncomfortable common ground. It is a known sociological phenomenon in group dynamics that reconciling differences is a more or less natural leveling process that generally seeks the lowest common denominator.

While it is a part of our American belief system that all men and women are created equal, it is immeasurably harmful to extend that notion further by suggesting that all ideas exist on equal footing as well. A society that cannot determine the relative merits of basic intellectual and philosophical principles is a society flirting with chaos.

The Myth of

Electronic Democracy:

A Reality Check

It is axiomatic in digital culture that the virtual world will give rise to new forms of electronic democracy. There is even a line of thinking that suggests that these new forms could be in some ways superior to our traditional form of representational government. Proponents argue that widespread use of computer bulletin boards will allow ordinary citizens a means of free expression previously unavailable under the oppressive shadow of institutional politics. In theory, this shining virtual city on a hill looks impressive and appealing—at least from a distance. But once again, a little poking and probing reveals that the reality is far more complicated.

Howard Rheingold makes a strong case for the democratizing effects of the Internet on society and culture. In *The Virtual Community*, Rheingold argues that online systems provide the best and shortest path around traditional forms of media, which have become overly centralized and distorted:

> Contemporary observers have documented and analyzed the way mass media . . . have "commoditized" the public sphere, substituting slick public relations for genuine debate and packaging both issues and candidates like other consumer products.
>
> The political significance of [computer networking] lies in its capacity to challenge the existing political hierarchy's monopoly on powerful communications media, and perhaps thus revitalize citizen-based democracy.[3]

There is certainly some substance to what Rheingold is saying. Concern over media concentration and monopoly seems all the more paramount given the signing into law of a new telecommunications reform bill in

February 1996. This legislation effectively nullified decades of work by the Federal Communications Commission to keep the broadcast media from being dominated by a relatively small group of owners. Media concentration is now a very real concern; megamergers continue to shape the new landscape of convergence between the publishing, computer, entertainment, telephone, and broadcast and cable television industries.

While the notion of electronic democracy is appealing on the surface, there are many hidden complexities involved. To understand them, we must move beyond the one-dimensional view prevalent in digital culture. Rigorous examination reveals a number of fallacies in that view. For starters, Rheingold and others make the huge assumption that online systems will be available to most of the American public within a reasonable period of time, and they assume these widely accessible systems will be universally taken advantage of.

These assumptions are at best questionable. By definition, a system, electronic or otherwise, cannot be democratic if not all of the populace is able to participate. At present, the number of Americans who are participating online is still only a fraction of the population as whole. Furthermore, few credible experts and analysts predict that this level of penetration will in the foreseeable future reach the majority of American households, at least given the workings of current market forces. The credibility of those who truly believe in the notion of digital democracy would be greatly enhanced if the rest of their agenda were more favorable toward those on the lower end of the economic ladder.

We have seen a few limited experiments with electronic democracy over the last five to ten years but little in terms of real implementation. A survey conducted by the Media Studies Center showed that only about one in ten voters used the Internet to access information about the 1996 presidential campaign, and fewer than 1 percent said that they relied primarily on that data in their voting decision. In what was perhaps the first known major attempt at electronic voting on a national level, Ross Perot's Reform party made use of the concept as a fundamental mechanism for voter feedback at its 1996 convention: The party appealed to voters to send a fax or an e-mail indicating whether they preferred Perot or his opponent, Richard Lamm. It is no coincidence that this approach was adopted by the first candidate for president who could legitimately be described as a technocrat. Perot, with his strong corporate credentials and his background in the computer industry, spoke frequently during the campaign about his unique qualifications. Among them was the ability to "manage complex systems," which he repeatedly stressed in the course of one speech.

Proponents of electronic democracy contend that it could allow a new form of government to emerge—direct, participatory democracy, as opposed to our current constitutionally based form of representative democracy. Interestingly enough, as radical as such notions appear at first blush, there have been serious and significant explorations of these possibilities at various levels of government and academia.

Many questions remain, however. Would such systems foster real democracy or simply lay the groundwork for a rather sophisticated means of manipulating public opinion? Would they erode workable democratic mechanisms now in place by subtly persuading people to exchange their right to vote for the right to float their opinions on any imaginable subject harmlessly into the ether? Since e-mail is notoriously easy to fake, what mechanism might be made available to make electronic voting invulnerable to fraud?

To some extent, a kind of electronic democracy has already been made possible by new communications and polling techniques that increase the speed and viability of direct information feedback. By contrast, although voting is itself a process of information feedback, it occurs at predictable and periodic intervals and involves extended deliberation.

The quickening of our collective information pulse that results from new digital technologies now means that the feedback process can be greatly accelerated. This acceleration offers the attractive quality—at least superficially—of bringing a greater sense of engagement into the political process. There are many dangers in this notion, however. As any musician who uses electronic equipment can tell you, too much feedback can result only in noise. Of far greater concern, however, is the possible weakening of due process and deliberation in the political system.

Yet another concern is the misuse of these new feedback techniques. Can opinion sampling be taken too far? A 1996 article in the *Wall Street Journal* described technologies that can track every movement of a person browsing the Web. Although the article downplayed it, the potential for abuse here is significant. In an even more troubling incident, political handlers during the 1996 presidential campaign used what are called "dial sessions" to ascertain voter reaction to various candidates' speeches. The technique works like this. Randomly chosen voters are provided with handheld transmitters to manipulate while they watch a preselected speech. On the transmitter is a dial that the participants turn to indicate their second-to-second reaction to what they are hearing. The entire group's reactions are then averaged by computer to provide a real-time interactive chart showing the group's collective response to the various twists and turns of the speech, as well as to gauge reaction to specific hot button subjects like crime, abortion, and social security.

Is this electronic democracy in action, or is it some cynical and bizarre form of political temperature taking? When we take into account the way in which this information is typically used, evidence points toward the latter. The essential purpose of the exercise is to see what kinds of managed reality and prepackaged political presentation are most effective with the voters on a purely emotional level. Second-to-second responses allow no time whatsoever for deliberation or the exercise of reasoned judgment.

It is difficult not to see this application of electronic democracy as anything other than cynical manipulation moved to the level of an art form. In fact, the pure-emotion driven political reactions produced by dial sessions are the raw material of manipulation.

In addition, there are distinct dangers in appealing to the electorate's emotional side over and above its rational and deliberative side. When politicians appeal to emotion on a purely instinctual level, without the benefit of such technology, they are called demagogues

Politics conducted by media handlers and politicians who are beholden to the whims of individual preference collected en masse is not politics but a sophisticated form of voter manipulation and pandering.

While some might be tempted to see this as a kind of electronic tapping of the will of the people, if wrongly implemented it could lead to reactionary politics or even to a kind of electronically enabled totalitarianism.

I do not mean to say that electronic feedback might not be useful if properly applied. The representative government that we now have was at one time a bold experiment in the history of nations. Perhaps the electronic plebiscite can carry this experiment even further.

There are no easy answers to these questions because we have no experiential precedent upon which to draw conclusions. Electronic poll taking is a fixture of the political environment and has been for many years. Its effects on the day-to-day political process draw largely mixed reviews. One school of thought suggests that we have allowed a class of politicians to take office that is slavishly wired to the whims of the populace, and that the notion of real leadership has taken a back seat to a finger-to-the-wind style of politics, predicated on opinion sampling and poll taking. Another school of thought, however, suggests that without the use of electronic media, there would be no practical and effective way of telling where the voting public stood on a variety of issues, aside from periodic, ad hoc special referenda. In this latter view, democracy is theorized to be functioning in perhaps a purer and less mediated form.

Certain analogies can be drawn between the casualness and informality of polling and the expression of opinions online. Both can be contrasted with more formal, deliberative, and ceremonial expressions of public will and opinion, such as the act of voting itself. Confusing the two can lead to problems. Where does posting in a newsgroup such as alt.politics, for example, stand relative to formal civic expression? Is it comparable to idle water-cooler conversation, or does it approach the bearing and stature of the more formal act of writing a letter to the editor?

Once again, contextualization is the key. A letter to the editor must be signed, and the writer must declare his or her identity and location as a matter of protocol. Can we say that posting an opinion online has the same political weight and importance? It would appear not, since rendering an anonymous or semi-anonymous opinion requires little in the way of self-declaration and correspondingly less personal risk. The very notion of political participation is predicated on citizens' willingness to stand up for their beliefs, and this act of self-declaration involves both courage and risk taking.

These notions of risk and self-declaration are important to any consideration of the complexities of electronic democracy. The act of writing was once described by Robert Frost as "risking spirit in substantiation." Like writing, voting is a substantive and ceremonial civic act that requires presence of mind and a certain amount of deliberation. Contrast the formality of writing and voting, then, with the offhand nature of both electronic polling and posting opinions on a bulletin board or conferencing system. Where is the larger sense of societal impact, ceremony, or civic gravitas?

Electronic polling courts and flatters citizens by asking them to render an opinion on a weighty and complex matter on which they probably have had little chance to form an intelligent opinion. No matter. Many people will gladly opt for this opportunity when given the chance, in no small measure because it confers the illusion that their opinion is being heard and that they are becoming more involved in the political process. But are they really? And could electronic democracy, as pleasing and simple a notion as it seems, ultimately undermine the sense of ceremonial seriousness that casting a ballot every two to four years can provide?

Perhaps the danger is that electronic democracy paradoxically creates more opportunities for expression in the public sphere while at the same time devaluing the significance of the individual messages. As we have seen, there may be an inverse relationship between quantity of information and quality of information. Although more opportunities for individual expression may exist, these expressions may, in the last analysis, count for less.

As exciting as these new tools are in terms of their ability to reinvent human activity and endeavor, the prospect of a technocratic power shift should not be taken lightly. And, like it or not, the Internet would certainly be a major factor in such a shift. The danger of technocracy does not lie only in the fact that the skills and worldviews of its proponents tend to be narrowly focused and lack the depth of perspective that our more humanistic traditions can afford. The more critical concern is that as the technological diaspora continues, the challenges of societal decision making will become irredeemably more complex while its locus of influence moves further away from the populace itself.

In a purely political sense, the problem of dealing with complexity is compounded not only by the nature of the subjects being considered but also by the methodologies that might be used to consider them. Such methodologies may, in the genuine attempt to respond to complexity, bypass the existing mechanisms that are the basis for our particular form of representative democracy. In other words, if society becomes too technologically complex, then decisions that lie at the intersection of technology and politics may also become too complex for ordinary citizens.

This possibility opens the door to a kind of upward referral and deference to the experts, who, in this case, are likely to be none other than the emerging technocratic elites under discussion. I suspect that digital culture's zealousness to replace traditional forms of government might mask an alternative possibility: that this new elite, under the smoke screen of electronic populism, could install itself as a substitute for the incumbent elite.

The not particularly pretty picture that then unfolds is that of a plebiscite mollified by the illusion of participation, electronically but largely anonymously chattering away about the issues of the day, having been persuaded in effect to exchange the singular power of the vote for the chance to participate in an electronic forum that has no true potential for deliberative action, to exchange real power for bread and circuses. In this scenario, real decision making continues, as it must, but its locus moves further away from the populace toward a technocratic elite whose control is derived from having a firm hand on the tiller of a technological machine that becomes an increasingly larger part of the political process.

Granted, this is a worst-case scenario. In due course, we will likely find that there are effective ways to bring electronic expression under the umbrella of political discourse. There may be ways, for example, to set up opinion polling in such a way that it avoids the lopped off, oversimplified thumbs-up-thumbs-down questions that are now standard in media-sponsored national opinion polls.

When the Internet is brought into the picture, dialogue replaces the polling process—which arguably is increasingly structured toward rather predictable outcomes—and becomes a real-time, interactive dynamic. In this sense, what is called disintermediation occurs, and the managing of public opinion that constitutes the polling process has, in theory, been replaced by access to open, direct, and free discussion. This anarchic pluralism (Mike Godwin has termed it "radical pluralism") is intriguing, particularly with respect to its implications for both new cultural and new governmental models.

While some critics have dismissed the teeming and multifarious discourse of the Net as little more than the formalized rant of the populace, others, like Godwin, point out that this utterly unprogrammatic and spontaneous immersion in the so-called marketplace of ideas amounts to a great social experiment that we should tamper with as little as possible. Godwin argues that the Net should be allowed to wander toward its own natural dialectical endpoints, that individual expression should be prized and cultivated, and that we should look at the result as a valuable product of this new means of human communication.

While this is an intriguing argument, it has more validity as a generality than as an axiom to be applied in the political realm. It remains at this juncture simply part of the reigning Internet mythology of electronic democracy and has yet to prove itself to be more than simply an idea— and perhaps an oversimplified and idealistic one at that—in the minds of some Internet commentators. In point of fact, Net discourse, despite many individual epiphanies, has remained largely disappointing and uncivil.

The idea of electronic democracy may be comforting to those who spend a disproportionate amount of time in the virtual world because that idea allows them to stoke the fires of their Net idealism and to feel that they are fulfilling rather than avoiding the practical complexities of the political involvement. But the fact remains that the Net is simply no substitute for the broader political activities of an actively engaged and knowledgeable citizenry.

A far more interesting prospect for electronic democracy is rarely discussed: local community use of online systems. When digital culture talks about electronic democracy and Net politics, the emphasis more often than not falls on the national and global reach of such systems. The conventional wisdom assures us that it is useful and interesting to get up close and personal with strangers who are hundreds or thousands of miles away. It even suggests that we are redefining the nature of community itself through online participation, another argument favored by Howard Rheingold. But, curiously, at a time when the integrity of local communities is besieged by a variety of social forces, we rarely hear about the prospect of using such systems to reinforce the cohesiveness of communities.

How could local bulletin board or conferencing systems overcome the forces of societal fragmentation rather than reinforce them? There are a number of possibilities. In the area of education, for example, conferencing systems could be established to discuss the quality of teaching at local schools, giving parents a real voice in the educational process, a voice less superficial than that afforded by participation in the PTA and other similar groups. Local Web-based conferencing systems might allow parents to have a say in the many problems confronting public school systems in towns and cities across the United States. The focus, however, need not be limited to education. Community support groups could be formed to help reintegrate communities that have been fragmented by a torrent of complex social and economic forces.

Will such systems ever get off the ground? Only time will tell. In theory, intranets, coupled with new software programs available through the Internet, could make it much easier for individuals or groups to set up ad hoc conferencing systems for just such purposes. Eventually, this capability will likely emerge, championed by some enterprising start-up company, and be adopted by savvy citizens. But to make such systems more feasible, things will have to change, including current network economic models and the cultural resistance of entrenched groups that perceive such grassroots input as unwarranted intrusion on their prerogatives.

For the sake of example, let us return to the case of a parents' conferencing system developed for the purpose of discussing the educational system. The first political and cultural issue that comes to the forefront here is sponsorship. If such systems were established by individuals or small groups rather than, say, the school board, they would still be viewed as private systems. As such, they might or might not exist with official sanction and participation. Community bulletin boards could allow parents to work together and make constructive comments for change, but there would be no formal mechanism for providing that feedback to a school administration.

The economic question is even trickier. If these systems were not initiated by the schools themselves, then who would pay for their operation? This question provides a good example of why complete privatization of network services is a poor model with respect to fulfilling the societal dimensions of emerging telecommunications systems. While solutions might be available, it is unlikely that the market in the natural course of things would provide them. New models of fee-for-service payback, whereby costs could easily be spread over a group or community, could provide the answer. However, no such models currently exist.

In general, our experience with community-based local networking is fairly limited, but I believe such networking is one of the most important and most underestimated potential benefits of the digital revolution. There have been a number of notable projects in this area, including the Santa Monica PEN project, Berkeley Community Memory, rural Montana's Big Sky Telegraph, and the Cleveland Free-Net system.

Unlike national conferencing systems, where the scattered quality of public discourse tends to preclude the possibility of real grassroots action and organization, local systems could negate the talk radio syndrome of the Net, in which comments rarely lead to constructive action. Local systems could offer a model for using online services to foster true electronic democracy by allowing discussion to lead to constructive action, rather than simply providing a means to vent frustration or float an opinion into the electronic agora like a message in a bottle. In this model, the Internet and other online systems could offer the potential for real grassroots political action. Furthermore, because they shift the locus of power from the federal and state governments to local governments, community-based online systems could facilitate the development of new social programs under this model.

In order for such scenarios to work, however, all community participants would need to have affordable access to online services. While flat-rate pricing with unlimited access is a major step in the right direction, there are no guarantees that such approaches will continue under market deregulation. We need only look at rising rates in the recently deregulated cable industry to get a sense of how this might devolve.

If electronic democracy is to have any meaning whatsoever, its proponents must make a convincing case that technology will not be used to separate and classify citizens and that it is fully compatible with the politics of inclusion. We would not think of holding an election if only 20 percent of the American people could vote. By the same token, electronic democracy will never succeed unless there is at least the reasonable prospect that online access will be nearly universal. Affordability, computer literacy, and sociocultural inclination are obviously key issues here

The sad reality is that in the current economic environment, even the cost of telephone use has shot out of reach for many Americans living on the socioeconomic margins. The increasing popularity of prepaid phone cards is, in part, due to the fact that for many on the lower rungs of the economic ladder the only telephone service available is a pay phone.

If basic telephone service has indeed become less affordable for many Americans, then what does that suggest about the future of electronic democracy? If the information revolution stalls out as it reaches the socioeconomic barriers of wealth and know-how, the online world will then truly represent a neo-Darwinian "great divide," separating the information-blessed from the information-deficient. Without appropriate vigilance, online systems could easily become more antidemocratic than democratic in nature.

Digital Culture

-------------- The most important magazine in the world today.
—*The London Observer*, in reference to *Wired*

-------------- If hyperbolic technological optimism were all that *Wired*
represented, it would be a modestly interesting, if sometimes
howlingly brainless, gazette of our times. Unfortunately, the
magazine and its popularity among the young, urban elite also
seems to represent something darker. *Wired*'s insider-outsider
dichotomy has the taint of contempt for the poor and the
uneducated, people not using computers now and not likely,
in the close future, to find a reason to use them. The dis-
advantaged haunt *Wired* by their absence, like a negative
space that can be seen but can't be accounted for.
—Gary Chapman

-------------- To be a god, at least a creative one, one must relinquish control.
—Kevin Kelly

-------------- Our contemporary shift from flesh to electronics is precisely a
shift from the human to the demonic.
—William Irwin Thompson

Why *Wired* Is Tired:

The Transformation of

Technology into Culture

The letter had arrived buried in a stack of press releases. Distracted as I was that day, I had almost missed it. But the striking signature, revealing a cosmopolitan, calligraphic style, caught my eye. The letter was from Nicholas Negroponte, the director of MIT's Media Lab, thanking me for sending him the most recent issue of our magazine, an issue for which he had penned an article.

Negroponte's article had found its way into *Telecommunications* by an interesting turn of events. The piece had been written as a rebuttal to another article in that same January 1992 issue, an article by Professor Joseph Pelton, who headed up the University of Colorado's telecommunications program. Its rather provocative title: "Why Nicholas Negroponte Is Wrong about the Future of Telecommunications." Pelton had been itching for a chance to puncture what he thought were Negroponte's overblown and inaccurate statements to the press about the future evolution of telecommunications networks. Here was his chance.

At Pelton's urging and in the interest of fairness, I had informed Negroponte about the upcoming piece and asked if he was interested in countering Pelton's arguments. Graciously, he agreed. We decided to create a special section to highlight the two articles, and I told our art director that I wanted it to have top billing on the cover. The publisher and I both agreed that the debate format would be an interesting one to present to our readers.

Negroponte's thank you letter mentioned in passing a publishing project he was involved in both as a senior columnist and an investor, a magazine called *Wired*. The letter talked a bit about *Wired*'s mission and suggested that, as the editor of *Telecommunications*, I might want to do a story about the new publication. He stressed that *Wired* was being designed almost as a counterpoint to traditional trade publications and that its focus would be on technology as lifestyle and not "pixels, bits, and megahertz."

The first issue of *Wired* hit the newsstands in January 1993. The look and feel of the magazine alone conveyed its uniqueness. This first issue featured a cover story by science-fiction writer Bruce Sterling on the use of digital technology in warfare entitled "War Is Virtual Hell," an interview with Camille Paglia contributed by Stewart Brand, a piece on cellular "phreaking" by *New York Times* reporter John Markoff, Negroponte's very first column (on high-definition television), an article by ex–*Whole Earth* editor Art Kleiner, and another by Gerard Van der Leun, the then director of communications at the Electronic Frontier Foundation (EFF).

Also included in that first issue was an introduction by then publisher and editor-in-chief Louis Rossetto, an effusive and hypermanic declaration, framed in the characteristic rich Day-Glo coloration and fifty-point type that would eventually become one of *Wired*'s trademarks. That declaration, *Wired*'s mission statement, offered lines like "the Digital Revolution is whipping through our lives like a Bengali typhoon—while the mainstream media is still groping for the snooze button" and "*Wired* is about the most important people on the planet today."

To give credit where credit is due, there is no question that *Wired* achieved significant commercial success in a relatively short order. Through a variety of techniques, many of them stylistic, the magazine was able to capitalize on an untapped opportunity and define a niche for itself as a publication that approached computer and communications technology from a radically different perspective than most trade and consumer magazines did. Rossetto said the magazine would achieve that different perspective by writing about the social impacts of the new technology. Interestingly, that lofty objective was somewhat at odds with the *Wired* that Negroponte had described to me in his letter, which had referred to *Wired*'s central motif as lifestyle. The distinction is not a trivial one, and in the last analysis it was Negroponte's description that rang true.

It is questionable whether the publication even came close to achieving Rossetto's goal of analyzing the social impacts of new technology. In this sense, *Wired* was a disappointment to anyone who expected to see Rossetto's promise fulfilled. The reason for this was fairly simple: The magazine failed to describe the impact of the Internet on society because its editorial perspective gravitated far more easily toward the narrow interests and closed-circuit world of digital culture—the group that considered itself the standard bearer of the so-called digital revolution.

Despite an unwillingness or an inability to tackle broader themes and fulfill its own charter, *Wired* did succeed in bringing the Internet into the mainstream while alerting the media at large to the fact that there was a lot more to the digital revolution than met the eye. Trade magazines that delved into the Internet—and there were many—had little chance of achieving the same goal simply because many were controlled-circulation publications; that is, publications not widely distributed on newsstands but rather designed to reach specific, mostly professional audiences within the computer and telecommunications industries. But many of them also failed to notice the unique cultural and lifestyle issues associated with the new technology—issues that became *Wired*'s stock-in-trade.

From the very beginning, *Wired* predicated its success on cutting-edge design, using dissociative and antiformulaic montages that, in McLuhanesque fashion, toggled easily between the text and image. The magazine, in fact, gave the more or less accurate impression that a media-centric battle royal was taking place between text and image, although there were those who felt that the latter won the day.

Critics charged that *Wired*'s approach represented the triumph of style over substance. But the magazine had succeeded in staking a claim and defining a new territory for itself, as evidenced by the surfeit of flattering stylistic imitations that emerged in its wake. Madison Avenue was agog with envy. Borrowing some of its trendy cachet from *Mondo 2000*, another cyberculture magazine, *Wired* used graphics that were, in fact, often stunning and impressive, and it relied heavily upon sophisticated state-of-the-art computer technology. A casual check of the masthead in any given month revealed that the magazine had almost as many graphic designers as editors.

In terms of technology coverage, executive editor Kevin Kelly was *Wired*'s "big thinker." Kelly had an admirable grasp of where the cutting edge of technological development was at any given moment. His instincts for the new and interesting and the various technological expressions of the "smaller, better, cheaper, faster" mantra of the computer industry were well honed. His expertise notwithstanding, Kelly and the rest of the editorial staff had, however, a technological Achilles' heel. While they understood computers, the Internet, and all of its associated technologies very well, they were nowhere near as knowledgeable about the complexities of the telecommunications industry and the nation's telephone system, which is the basic infrastructure on which all Internet communications rest.

Many of *Wired*'s articles were drawn from outside contributors rather than written by the editorial staff. By adopting this sourcing strategy, the magazine managed to showcase some of the better commentators within the computer community and at its fringes, where digital culture was newly emerging. However, the magazine only rarely called upon experts within the telecommunications industry—a practice that betrayed a certain lack of technical balance and sophistication. In the publication's early days under Kevin Kelly and managing editor John Battelle, *Wired* gradually built up a stable of accomplished freelance contributors, many of whom were also WELL habitués and seasoned cybernauts in their own right: Steven Levy, who went on to become a technology writer for *Newsweek*, Paulina Borsook, Mitchell Kapor, Howard Rheingold, Jon Katz, Brock Meeks, John Markoff, and many others. Within the narrowly defined parameters of their selected subject matter, the quality of writing and reporting was often first-rate.

In 1994 and 1995, the Internet began to garner even more media attention. During this time, the burden of explaining the implications of the new cyberspace paradigm increasingly fell on a group of early adopters and power users of the Internet—the bright, knowledgeable, well-educated, and for the most part rather privileged members of the digital elite or, more colloquially, the digerati. Lotus Development Corporation founder, former transcendental meditation teacher, and gnomic whiz kid in CEO guise, Mitchell Kapor seemed to be at the center of much of this activity.

Kapor had made his millions, and now he had a new goal to pursue, one that might even be described as idealistic. He would spend his time creating a right-thinking environment for the new and quite possibly politically fragile Internet technologies to flourish in. With Net politics as his newfound hobby, he teamed up with former Grateful Dead lyricist and self-proclaimed cyberpoet John Perry Barlow to found the EFF in 1990. One of the main tasks of the EFF would be to foster understanding among both Washington policymakers and the general public concerning the importance of not constraining the growth of cyberspace with the laws, customs, and mores of the nonvirtual world. About a year after the EFF's founding, its board consisted of Stewart Brand, John Gilmore, Barlow, Esther Dyson, Kapor himself, and several others. Many of these EFF insiders also had strong ties to *Wired*'s community of interest.

With the founding of the EFF, the digital elite seemed to find a new center of gravity. As the EFF became more politically active and helped to define some of the key concerns that would emerge when cyberspace itself became a policy issue in Washington, it became evident the EFF was one of the few entities that possessed substantial insight into what the Internet was really all about and how it was going to affect at least some aspects of society and culture. However, Kapor and the EFF were not so much original thinkers as they were popularizers. To a certain extent, they picked the best of Net wisdom, packaged it in sound bites that the media and Washington policymakers could readily understand, and used this information as the basis for their missionary work.

When *Wired* arrived on the scene, the digital elite found yet another vehicle. What is more, a certain resonance between Kapor's EFF and *Wired* soon became evident. However, this resonance was hardly surprising, given the amount of philosophical and organizational commonalities that existed between the two entities. For one thing, Kevin Kelly, executive editor of *Wired,* was a protégé of Stewart Brand, one of the early EFF board members. And many EFF staffers penned articles in *Wired* that reflected a kind of emerging unified field theory of the Internet. A number of EFF representatives made it onto the masthead as contributing writers or in some other capacity. This harmonic convergence reached its height when *Wired* published a sycophantic article by journalist Joshua Quittner that apotheosized the EFF and revealed that the organization was considering becoming a political party.

As *Wired* became increasingly successful as both a newsstand and a subscription-based publication, the mainstream media started taking notice. Breathlessly enthusiastic articles declared *Wired* to be incontrovertibly cool and upscale—this despite the fact that the writers generating these articles were hardly in a position to make accurate assessments of the magazine's content. As with all trends, you either signed on or you did not, and those who joined the media juggernaut did not always do so for the soundest of reasons. The mainstream press sensed that there was a richly exploitable pop-culture angle to the magazine's hazy effusions, insider tech-talk, and oblique Emersonian assertions about the future direction of the Net. No doubt motivated in part by the media's need to shore up their sagging Generation X readership, this kind of coverage seemed like the perfect solution, and a new symbiosis was born.

Negroponte's goal for *Wired*—that it become a publication not just about technology but about lifestyle—was fulfilled. Reporters who were not Internet savvy gravitated to *Wired* and its coterie of writers, editors, and techno-groupies because they seemed to "grok" what the digital revolution was all about. It was also helpful that they had little hesitation about sharing their views with anyone who would listen. It was a cause célèbre, a mission of the highest order, a vision of a new society for those few individuals who possessed the requisite technological savoir faire to fully appreciate the significance of the New Media's impact.

No question about it, *Wired* seemed to have it all figured out. There was even an electronic hangout (actually there were several) where the digerati, including many of the more visible pundits and opinion makers, gathered to compare notes and reinforce the particular mythologies being forged by this unique fusion of technology and popular culture. One important venue was the WELL, a virtual pit stop for many of the thought leaders of cyberculture, including Howard Rheingold, Kevin Kelly, Mitchell Kapor, and John Barlow.

It is difficult to say exactly how and when digital culture surfaced as a discernible phenomenon. The year 1993, when the Internet first attracted the media's attention and *Wired* was launched, certainly seems to be a handy reference point. Whatever the historical trajectory, *Wired* and the EFF were at the white-hot center of the new technological diaspora.

The metamessages in *Wired* were clear and unequivocal: The digital revolution was here to stay, it was unstoppable, and left to its own devices without interference from government it would bring about some sort of new world order. *Wired* would be the serialized equivalent of Martin Luther's ninety-five theses nailed to the door of All Saints' Church in Wittenberg, Germany. But *Wired*'s insistent subtext was not a declaration of war so much as a declaration of inevitability.

Wired's proselytizing stance took a lot for granted on the part of its readers. For one thing, the magazine seemed to expect its readers to share its assumptions about the Internet as a prerequisite to "joining the club." Many of these assumptions were so general that it was difficult to argue with them. How can you argue with a graphically striking, highly stylized pastiche that embeds cryptic quotes from Marshall McLuhan in thirty-point type?

The nebulous cloud of assumptions that pervaded *Wired*'s message tended to belie the notion that science and technology—with their natural propensity for order, accuracy, and precision—were involved here. No, there was something else present, a leap of faith perhaps, or possibly a kind of cultural mutation that operated conveniently out of the range of the expected set of journalistic descriptive powers. Many of the articles in *Wired* were less articles than encrypted messages intended for the initiated. The writers knew what was going on, and they were going to let you in on it, provided, of course, that you did not mind continually playing catch-up and did not ask too many unseemly and annoying questions. But for those whose wanted more clarity, this particular kind of ignorance could certainly not be confused with bliss.

In terms of useful content, *Wired* always seemed to be five steps ahead of its readers, and that, of course, was exactly the effect the magazine wanted to achieve. In terms of its reliance on overarching assumptions, it usually managed to avoid precisely defining the specific benefits that would accrue to society at large as a result of the digital revolution. It was much easier (and probably more effective) to provide the textual equivalent of French impressionism by employing evocative imagery and summoning a quasi-mystical sense of the evolutionary power of computers and communications to transform human existence.

As a part of its relentless self-promotion as the magazine of record for the digital revolution, *Wired* was fond of claiming that digital culture was carrying on the legacy of the sixties. Superficially at least, in terms of youthful energy and spirit, the comparison seemed valid. The magazine did have a certain amount of countercultural cachet, which was applied liberally to back up that claim. For example, executive editor Kevin Kelly had formerly been editor of *Whole Earth Review,* a magazine that could indeed trace its lineage back to the sixties and the cultural icon known as the *Whole Earth Catalog,* founded by Stewart Brand. Besides Brand and Kelly, there were other ex–*Whole Earth*ers who ended up in *Wired*'s orbit either as writers or contributing editors.

In the sixties the alternative press sprang up to provide points of view that could not be adequately reflected in mainstream venues, but despite the *Wired*'s self-professed disdain for media empires, *Wired* and Big Media seemed to be made for each other, for a symbiotic relationship. Unlike *Mondo 2000*, which also had pretensions to being countercultural, *Wired* was almost a kind of *Fortune* magazine for hip, young, largely corporation-dependent digital professionals, whose success was deeply intertwined with that of the computer industry. For them, the idea of dropping out sixties-style was tantamount to excommunication from the electronic polity. Both magazines could be seen as crucibles for a new kind of technoid lifestyle, which, I will argue, laid the foundation for a new technocratic set of values that gained mindshare not on the basis of being rigorously examined and then accepted but rather unquestioningly swallowed whole on the basis of its trendy cachet.

Wired's position continued to be enhanced by its status as a darling of the media, which at one point declared the magazine to be the new arbiter of all things digital. In due course, the same names kept popping up in the same contexts: Kapor, Kelly, Barlow, Brand, Dyson, Rheingold, Sterling, and of course Negroponte, all card-carrying members of the digerati.

In general, these media commentators showed little interest in how the Internet related to U.S. telecommunications infrastructure and policy. Either the digerati did not particularly care about providing this larger picture, or they were not particularly good at it. But there was one other curiosity that emerged: With respect to the Internet and its effects on society and culture, most of the digerati appeared to be singing from the same hymnal. The uniformity in their assumptions about the social value of the digital revolution was rather striking in such a group of seemingly independent thinkers.

As interesting, knowledgeable, and up-to-speed on technological issues as they were, their views were surprisingly predictable and insular. For example, with the exception of Howard Rheingold and a few others, none seemed willing to concede even the remotest possibility that negative effects or unintended consequences could accrue as a result of the digital revolution. None addressed the tough questions. Although it was generally accepted that the Internet would bring about a sea change in the way that most Americans would experience the important functions of business, education, and health care, there seemed to be a surprising lack of willingness to explore exactly what its specific impacts might be. Faith prevailed over reason.

In general, *Wired*'s editorial content was largely Internet- and computer-centric. Favorite themes included many of the quasi-political issues that the EFF was involved with, such as the Digital Telephony Bill, the infamous Clipper Chip crusade, and the Communications Decency Act. Encryption was always a hot topic, almost an obsession.

On the subject of government regulation, *Wired* invariably reinforced hard-core libertarian notions: total market deregulation and the abolition of universal service, the system of subsidies that for decades had ensured that all Americans could enjoy basic telephone service at reasonable rates. If *Wired* had any particular concerns about the ongoing partitioning of the America public into the information-rich and information-poor, those concerns were well hidden, except in the occasional contrarian article that crept in from time to time. In short, the consistent focus of the magazine clearly and specifically addressed the needs of digital culture rather than the needs of most Americans with respect to the influence of new technology. Behind the magazine's superficial considerations of lifestyle, a slew of significant assumptions about the nature of contemporary life and so-called digital politics slumbered uneasily.

In addition to the longer and more thoughtful trend articles, there were short pieces on the latest and greatest techno-gadgetry; reviews of upscale products; more than a few fawning interviews with various captains of industry such as John Malone, president of cable giant TCI, and Ray Smith, former CEO of Bell Atlantic; gossip and snippets about computer industry mavens and their rising and falling fortunes; and much editorializing, alternatively offhand and passionate, on the subject of free speech on the Internet.

From a technical standpoint, encryption, virtual reality, computer warfare, the Web, digital politics, free market deregulation, nanotechnology, and computer games were standard fare. Occasionally, a reader might find a more "far-out" article, invariably delivered with a surfeit of enthusiasm celebrating the "technological sublime," what *Wired* seemed to think were the more "mystical" aspects of digital technology. In a few cases, the phenomenon of the Internet and the evolutionary theories of Catholic theologian Pierre Teilhard de Chardin were linked in an overt attempt to connect the technological with the spiritual. One of the masters of this particular genre was John Barlow, once described by Kevin Kelly as a "hippy mystic" (hippy Republican mystic might be more accurate). From time to time, there were, of course, articles that reinforced *Wired*'s clever marketing inversions about the sixties—articles implying that, just by sitting at their keyboards, screen jockeys were making the same kind of commitment to social change that counterculture idealists had made.

In the final analysis, what *Wired* did not talk about was more revealing than what it did talk about. It did not, for example, deign to report on major technical or marketing trends in the telecommunications industry or important developments in the shaping of national communications policy (unless, of course, those events happened to fall into the category of encryption, free speech, or related themes). Despite Rossetto's ringing proclamation in the premiere issue, the magazine tended to be surprisingly uninterested in providing any deeper analysis of the social impacts of the Internet and its associated technologies. There was a section with the somewhat presumptuous title of "Idées Fortes," a banner under which short and often obscure guest editorials appeared; but even here the subject matter was severely constricted in terms of the broader trends that were shaping the telecommunications landscape. It was almost as if *Wired*—with its prolibertarian, antigovernment bias—was intent on giving its readers permission to disengage from the larger arena of communications policy, a real-world arena in which social objectives were merged with the application of the technologies themselves; this was an ironic stance for a magazine claiming to be at the forefront of explicating the digital revolution's social impacts.

Clearly, the broad focus, diversity of opinion, and well-groomed cultural ecumenism that had been a defining characteristic of *Whole Earth Review* under Brand and Kelly was now gone. In its place was a carefully constructed worldview that had fixed borders, even if those borders were blurred by gratuitous flashes of pseudopsychedelic high-resolution visuals. *Wired*'s own narrowness of vision was cleverly belied by its continual invocation of infinite possibility. As Paulina Borsook, a former contributing writer at the magazine and one of the more astute observers of the undercurrents of digital culture, put it: "Although *Wired* maintains a posture of celebrating the cacophony and all the lack of prior restraint Net culture has to offer, in fact, it is not open to points of view other than its own, as bounded a set as *The National Review* or *The Advocate*."[1] At *Wired*, executive editor Kevin Kelly was no longer constrained by the mandate to foster the intellectually wide-ranging, independent inquiry that had been the hallmark of *Whole Earth* publications. He was now free to shape a partisan and narrowly focused magazine that could elegantly appropriate *Whole Earth*'s counterculture cachet while at the same time systematically undermining many of its principles.

As *Wired* continued to refine its editorial niche, the incestuous narrowness of its closed-circuit world began to take its toll. Editorial material began to look increasingly recycled. As much as *Wired* liked to claim that it was boldly going where no cybernaut had gone before and charting new paths into the electronic frontier, habitual themes were reinforced and repeated in issue after issue, with different packaging and different titles and bylines. *Wired* had defined who was with the program and who was not, and ideas that clashed with its own particular set of assumptions had little chance of getting published.

One of the most devastating critiques of *Wired* and its editorial tunnel vision came from Gary Chapman, director of the 21st Century Project at the University of Texas and former national director of Computer Professionals for Social Responsibility (CPSR). In the January 9, 1995, issue of *The New Republic,* Chapman wrote:

> *Wired* is significant not only because of the buzz it has created. It is the first successful publication to address the social and cultural effects of digital information technologies, as opposed to the technical details of machines and software. Clearly these technologies are reshaping society and they have become central features of many daily lives. But this worthy promise for a magazine has been spoiled by *Wired's* fevered, adolescent consumerism, its proud display of empty thoughts from a parade of smoke-shoveling celebrity pundits, its smug disengagement from the thorny problems facing postindustrial societies, and, most annoyingly, its over-the-top narcissism. If this is the revolution, do we really want to be part of it?

Wired's tendency to look out for the narrow interests of the emerging digital elite resonated well with the EFF's equally constricted social and political agenda. Although CPSR, like the EFF, traditionally worked on a variety of public policy issues pertaining to the emergence of the Internet, it, unlike the EFF and *Wired,* consisted of a broad range of socially aware professionals throughout the computer community who were working toward a positive and comprehensive social vision for the democratization of cyberspace. The agendas, stated and unstated, of both *Wired* and the EFF were far more elitist than CPSR's and were geared toward the concerns of their small but increasingly influential constituencies. Both entities were almost obsessively concerned with security and privacy issues. The locus of their concern was not, however, the potential for corporate abuse but rather government initiatives such as the Clipper Chip and the Digital Telephony Bill. Paulina Borsook wrote, "Technolibertarians rightfully worry about Big Bad Government yet think commerce unfettered can create all things bright and beautiful—and so they disregard the real invader of privacy: Corporate America seeking ever better ways to exploit the Net, to sell databases of consumer purchases and preferences, to track potential customers however it can."[2]

While issues such as the Clipper Chip were, in fact, legitimate cause for concern, they represented only a small fraction of the public policy issues the new digital technologies brought into the spotlight. Others included such concerns as equitable pricing and access for online resources; educational policies for moving technology into the classroom; the types of new models that might be put in place for universal service or its equivalent in cyberspace; the viability of electronic democracy and the most effective way to implement it; and the way that communications capabilities could be fairly and equitably extended throughout society.

It is not that privacy, encryption, and free speech are unimportant. If at some point in the future, a large number of Americans use the Net to conduct business, educate themselves, make purchases, and enjoy its recreational dimensions, then they ought to be able to do so with the reasonable assurance of privacy and the lack of government intrusion. It is just that in their all-encompassing zeal to deal with these issues, both *Wired* and the EFF seemed to have forgotten that there were many other important issues surrounding the use of cyberspace that needed attention and championing. Their obsession with privacy left the door open for the criticism that digital culture was really most concerned (and rather obsessively so) with its own narrow agenda rather than with important matters of public policy that would inevitably affect much wider constituencies.

Random Thoughts
on the Defining Works
of Cyberculture

How seriously should we take the ruminations of cyberculture, which seem to veer between the "insanely great" and the inane and all shades in between? As the Internet continues to expand both as a physical communications system and as a dimension of societal expression, clues about the future effects of these technologies on our collective imagination need to be gathered where they are found—in their natural state. There is much of value in digital culture: There are strong and worthy voices, there is keen intelligence and even prescience. But there are some ominous pointers toward less-than-desirable futures as well. The lack of academic rigor in some of cyberculture's formulations suggests not that its offerings should be devalued or discarded but that they should be taken on their own terms and with a different perspective and perhaps even a sense of playfulness that allows for new ways of looking at things.

It is still far too early, I believe, to establish a worthy and workable canon of books about the Net—too early to separate what will last from what attracts our fleeting attention with flashy chimeras of the technological sublime and the marketing blandishments of publishers who sometimes barely understand what they are publishing. But the effort to establish and organize the progression of thought in the existing literature is certainly a worthwhile exercise. Here I will attempt only a rough approximation in the interest of beginning a process that many others will refine as the Net continues to develop its contemporary mythologies and cultural mazeways.

Much of cyberculture thinking to date has been crystallized by various thought leaders in books and articles that address the new technological transformations quietly rearranging the landscape. These opinion shapers include many whose names we have already encountered: Kevin Kelly,

Stewart Brand, Howard Rheingold, Esther Dyson, and Nicholas Negroponte, all of whom have written books defining various themes associated with the emergence of the virtual world. In his or her own way, each
has made worthy and interesting contributions to further defining the
parameters of a heretofore unwritten manifesto of digital culture, a new
way of thinking about computers and a mapping of how they might be
expected to evolve technologically.

Steward Brand's book *The Media Lab: Inventing the Future at MIT*
was characteristically ahead of the curve when it appeared in 1987. Well-
written and highly readable, it is the result of a stretch of time during
which Brand, well on his way to becoming a true believer, hung out at MIT's
famous computer technology think tank, the Media Lab, as an observer.
The book consists largely of Brand's firsthand reports on the wide range
of projects that the Media Lab was working on with the help of funding
from a large and ever growing number of corporate sponsors. The book discusses the pioneering work being done by project leaders like David Zeltzer,
Stephen Benton, William Schreiber, and Alan Kay, each making his own
particular contributions to the emergence of the New Media.

Interpenetrating the book is Brand's unquestioning certainty that
digital technology will transform our way of life whether we like it or
not. As Brand puts it, "Once a new technology rolls over you, if you're not
part of the steamroller, you're part of the road." Curiously and reveal-
ingly, Brand's own perception is that the Media Lab existed in part to
"use computer technology to personalize and deeply humanize absolutely
everything."[3]

Another major contribution to the newly emerging landscape of
digital culture came from a Brand protégé, Kevin Kelly. In the hiatus
between his long stint as editor of *Whole Earth Review* and joining *Wired*
as executive editor in 1992, Kelly wrote *Out of Control: The Rise of Neo-
biological Civilization,* a massive text of more than three hundred pages
with a detailed, annotated bibliography. Published in 1994, the book purports to describe many of the more interesting developments taking place
in the realm of advanced computer technology under the overarching
notion that our technological and societal future will eventually be pred-

icated on a "marriage of the born and the made," a kind of hybridization of machine life and biological life (hence, the term "neo-biological"). Interestingly, Kelly's ruminations, many of them anecdotal and based upon interviews with primary researchers, deal only obliquely with the Internet and computer networking.

According to some knowledgeable observers, Kelly's book is a kind of bible of digital culture and provides a passable intellectual framework for some of that culture's less methodical assumptions, to put it politely. It is full of arcane concepts such as "swarm systems" and "hive mind," which, uncharacteristically in the traditions of scientific exposition, are never completely defined. Like *Wired,* the book contains a surfeit of evocation without explanation, of allusion without exegesis.

Out of Control is ambitious in scope and theme, but as a treatment of evolutionary direction it fails to convince because of its rather narrow vision of human spirit and enterprise. The book offers a spirited but indirect argument for man/machine hybridization and, in the final analysis, is a strong apologia for genetic engineering, the adoption of hypertechnologies (such as nanotechnology), the creation of computer-based artificial life, and the application of virtual reality. Predictably enough, like his counterparts, Kelly never seems interested in pondering the deeper, more humanistic questions associated with these technologies.

It is not an uninteresting book by any means, and Kelly's bright intellect and tight writing style make it a pleasure to read. However, as a Magna Carta for digital culture, as a grandiose vision of neobiological civilization, it is not particularly disciplined. The first few chapters lay out in broad terms the marriage of the born and the made that Kelly envisions, and the notion resurfaces at convenient narrative junctures. However, Kelly does little in the way of effectively envisioning what kind of society we might be building with these still untested tools of computational biology—a potent form of bioengineering—and other next-generation hypertechnologies being developed at places like MIT and Los Alamos National Laboratory. He begins where all scientists begin—with a tabula rasa, a "greenfield" palette, wholly receptive to a new regime disconnected by design from what preceded it.

Out of Control is certainly intellectually challenging, and it puts certain aspects of the digital revolution on high ground. But its defect is that it is not compellingly thoughtful or properly reflective about the powerful ideas it purports to discuss. Kelly rarely introduces the slightest bit of existential questioning about the social and moral implications of his themes. The attack of angst that flickers across Kelly's brow at the end of the chapter entitled "God Games" is only a mild one. Here Kelly muses on the Old Testament story of the Creation and the metaphysical implications of creatures becoming creators:

> Will creating our own simulacra *complete* Yahweh's genesis in an act of true flattery? Or does it commence mankind's demise in the most foolish audacity.
>
> Is the work of the model-making-its-own-model a sacrament or a blasphemy?[4]

In evaluating the book and Kelly's role as one of the thought leaders of the digital revolution, we have to decide whether to accept his central thesis that the modalities of the new computational biology are really about eschewing control. Let us consider for a moment that the opposite is true and ask whether, if allowed to run out of control, these new hypertechnologies might, in fact, become a means of establishing a special kind of control of both nature and society. The mastery of natural forces has been, of course, a primary impulse throughout the history of Western scientific thought, and to some extent it has been reinforced in a religious and historical context by a system of values found in Christianity. In Christian terms, a central argument tends to focus on whether God granted humankind the "stewardship" of nature so that we could have dominion over it and use it for our own purposes. Is it a mere accident then that Kelly happens to be both a fundamentalist Christian and a staunch advocate of biotechnology?

The real question is this: Are Kelly, and fringe scientists like Hans Moravec, really interested in relinquishing control? Or is Kelly's apologia for bioengineered and artificially developed systems yet another Orwellian inversion that represents a last willful attempt on the part of scientific materialism to hold onto its hegemony in the face of a new and more humanistic paradigm?

It seems clear that technologists who are trying to discover—and I prefer not to pull any punches semantically here—the secret of life and use that information for corporate and commercial gain are most certainly in the business of control. The quest to develop artificial life, in fact, represents an impulse to manipulate the very essence of life itself, DNA, and to profoundly alter the earth's ecological profile under the mantle of taking human evolution to the next level—an ideological praxis involving science's ultimate conquest of nature and the denouement of centuries of struggle for supremacy over the constraints of our biological heritage. Stewart Brand's dictum "we are as gods" is easily invoked here.[5]

One thing seems abundantly clear: If an increasingly hyper-technological and metaphysically threatened scientific establishment needed a resonant and high-minded justification for its continuing interference with the natural order under the rubric of progress, then Kelly's manifesto and the Orwellian inversions of digital culture could provide it. Any real sense of letting go and eschewing control would have far more to do with a return to a basic respect for natural processes cultivated by ancient cultures whose value systems, though long suppressed, continue to poke through the chinks in the rusting armor of scientific materialism. It is perhaps also no accident that it is in the laboratories of New Mexico, where the most recent great and world-altering scientific discoveries took place with the harnessing of the atom, that we are poised once again to unleash still more powerful forces, even though we have failed, from the standpoint of human wisdom, to master its effects. In Arctic regions and elsewhere, improperly contained radioactive waste continues to leak into the ocean.

Another major shaper of the emerging vision of digital culture is Howard Rheingold, a technology writer who served for a time as editor of *Whole Earth Review*. Rheingold is another member of Brand's coterie and an erstwhile contributor to *Wired*. He was also a highly visible presence on the WELL. Rheingold tends to offer a bit more humanistic balance to the starker visions of machine worship offered by Kelly and others. Rheingold's book *The Virtual Community* focuses on the online dynamics of the

WELL, the Internet, and other computer conferencing systems. In general, he tends to focus on human communities of interest on the Net and other systems while relentlessly stressing the superiority of virtual modalities.

Other books that have contributed strongly to the mythos of digital culture, albeit less directly, are Negroponte's *Being Digital* and Esther Dyson's *Release 2.0*. Mark Dery provides a helpful panorama of the hybridization of the technological and cultural in *Escape Velocity,* although his approach tends to be more descriptive than evaluative. And Thomas Mandel and Gerard Van der Leun have provided a useful compendium of online mores in *Rules of the Net,* which, as much as any other work, gives a sense that a distinct Net culture really exists.

In addition to providing the thought lineage of the so-called digital revolution, all of these works have contributed to defining the parameters of Net culture as a distinct and describable phenomenon with its own set of assumptions, customs, and social practices. In many respects, they shape and define the conventional wisdom (in the best sense of the term) and reflect their authors' abilities to grasp and express in readable fashion many of the more obvious and immediate implications of computer and communications technology for society at large. Where they fall short is in offering any coherent vision of what it would actually be like to live in a society that embraced these technologies with the authors' enthusiasm and unquestioning acceptance. At night, the earth's cities are shining necklaces of light when viewed from thirty-thousand feet. At street level, the perspective is somewhat different.

It is interesting to note that, with the exception of Kelly, none of these commentators purports to offer truly original thinking in the conventional academic sense. And even Kelly, in the acknowledgments to *Out of Control,* perhaps too modestly demurs by claiming: "Hardly an idea in this volume is mine alone." In contrast, both Brand's and Rheingold's books represent a kind of narrative reportage with some measure of analysis. Even Negroponte—and keep in mind that he is an MIT professor—adopts the tone of narrating and describing an inexorable set of trends rather than offering a cohesive set of theories about the digital revolution. There is always the vague and unsettling sense of fait accompli, a sense that these books are postevent announcements of the arrival of some impressive but predetermined technological landscape that must be welcomed with open arms.

From a purely analytical standpoint, the almost anti-academic nature of these books begs for further explanation. That the earliest works defining what many would consider to be the cutting edge of the major transformation in both business and culture that we call the information age would come from a maverick group of independent writers and researchers is in and of itself worthy of exploration. However, it would appear that some measure of groupthink is also at work, since these individuals are bound by common involvement in the WELL and a special set of work associations centering on Brand's *Whole Earth* organization, Negroponte's Media Lab, and Kapor's EFF.

It is doubtful that either Brand or Rheingold would describe himself strictly as a journalist, and yet what they undertook were clearly journalistic efforts. Kelly calls himself a journalist but writes and thinks like a scientist. Negroponte stands apart, a university professor who freely admits that he does not like to write or read, and he considers the Media Lab's "demo or die" philosophy to be the functional equivalent of publish or perish. Writing and scholarship appear to be a kind of necessary afterthought rather than the essence of one's focus and activity. With its implicit emphasis on getting the product out the door and worrying about the details later, this philosophy might be compatible with the objectives of the Media Lab's long list of corporate sponsors. Like astronomer Carl Sagan, Negroponte is a popularizer. However, whereas Sagan was a well-respected scientist, Negroponte seems to represent a new breed of technologist who is as comfortable grandstanding on CNN as he is working in the lab.

In general, the books I have mentioned are loosely structured reportage and lack the systematic rigor that one associates with the practice of scientific discipline. Brand, Rheingold, and Negroponte adopt the stance of journalists as they try to diffuse and popularize the core values of digital culture, but in the final analysis, Kelly's book is the most intellectually challenging in that it deals with the frontiers of computer science and pushes the envelope of ideation. It is also worth noting that

Rheingold's book is the only one that deals extensively with the Internet and the nature of online communications, but the timing of Internet commercialization may be a factor and mitigating circumstance.

Interestingly, an outside observer who took at face value the breathless declaration of the EFF's John Barlow that the Internet is the greatest invention since "the capture of fire" would be surprised at the relatively small number of thorough and well-documented treatments of the societal impact of the digital revolution. At first blush, there appears to be a kind of "cognitive dissonance" at work. This dissonance could be partially explained by the fact that the Internet, as a subject or theme, seems to occupy a curious middle ground between computer science and pop culture. Something about the Net propels it outside the domain of pure science and technology and into a curious mix of related areas such as the media and entertainment. "Convergence" can certainly partially explain this phenomenon, but there seems to be more going on as well.

Which academic disciplines can claim the Net is still very much an open question, since the Net crosses many boundaries in the breadth of its impact. As a scientific phenomenon, apart from the tangential areas described in Kelly's book, the Internet has yet to become a formal object of academic study. That the *New York Public Library Science Desk Reference,* published in 1995, barely mentions the Internet (or telecommunications for that matter) is both puzzling and instructive. In fairness, however, it should be pointed out that we are still only beginning to evaluate the role that the Internet will play in shaping the mainstream social, cultural, and business agendas of the twenty-first century.

The books cited so far are not the only attempts to gauge the impact of the information age. But I wish to draw an important distinction between the specific set of ideas and assumptions that are associated with digital culture and those that are focused on the more generalized issues associated with emerging computer and communications technology. Although it is in the thinking of digital culture that we are likely to see the best and worst of what the digital revolution, with all of its complexity, might eventually have to offer, there are many other treatments of the information age available. Those worth noting include, for example, Alvin

Toffler's insightful *Powershift*. And there is an ample supply of books, from authors like George Gilder and Don Tapscott, that deal with the impact of computer networking on emerging business paradigms. (It is interesting to note that Kevin Kelly has joined the ranks of the latter with his most recent book, *New Rules for the New Economy*.)

It seems important to reinforce the idea that the thought leaders of digital culture understand the technological dimensions of the digital revolution extremely well, even if their analysis of societal impacts is either non-existent, narrowly focused on mega-issues like security and privacy, or myopic and biased on the side of technology for technology's sake. Their perspective makes the quality of their commentary on the technology itself worth paying attention to. The thought leaders of digital culture were also early adopters of the Internet and accomplished power users. They based their observations on practical experience rather than abstract theorizing. The Internet is an experiential phenomenon, where, to invert an academic axiom, those who can, teach, and those who can't, don't.

The New Media:

Tossing Out the Rules

What industry analysts like to call convergence—by all accounts an overused and abused term—involves the merging of computer, telecommunications, entertainment, and publishing technologies and markets. Needless to say, the new media environment that convergence makes possible will depend upon new ground rules for how information should be shaped, processed, disseminated, and interpreted. But as Everette Dennis, the director of the Media Studies Center, has warned, these new information systems are being designed and developed in many cases by a new set of organizational owners and managers who have responsibility for both content creation and content distribution. These new managers seem to have little appreciation for the traditional rules for ethical and responsible presentation of news and information to mass audiences. (This is not to suggest, of course, that established media providers are not also having difficulties in this area.)

The abandonment of the traditional goal of journalistic objectivity is just one example of this lack of appreciation. I was quite surprised, for example, to find how little regard for objectivity there was among the more cybersavvy journalists who frequented the WELL's media conference. But the anarchic exuberance and postmodern second-guessing that characterize the online world have now spilled into print. In many cases, particularly where online publications are concerned, New Media publishers and editors are, in effect, tossing out the rules of traditional journalism.

Why should this be the case? Digital culture justifies this practice by claiming that such rules are part of the elitist, institutionalized old order and, therefore, are somehow tainted. Under the rubric of such fanciful constructs as "Way New Journalism," concocted by *Wired* contributors like Brock Meeks, articles are developed on the basis of the whims of the writer rather than guided by a set of journalistic principles. (Blame the observer-dependent New Journalism of the sixties if you like.) Such constructs are related to the principle of "chaotic" communication, one of the fuzzier ideals of Internet culture.

Two of the four converging media industries—telecommunications and publishing—have traditionally relied upon a well-developed system of values and norms for the shaping and distribution of content. The telecommunications industry, as described earlier, has a rich history of normative public policy that routinely gets refined not only by government agencies like the Federal Communications Commission but also by university communities and policy think tanks like the Center for Strategic and International Studies or the Annenberg School. The field of publishing also has long-standing rules and guidelines that responsible journalists are enjoined to heed by their employers as well as by their own professional and peer group organizations.

In the case of the entertainment and computer industries, however, no such strictures apply. Given the blurring of roles that convergence both creates and allows, Microsoft has now become both a publisher and a broadcaster. MSNBC represents its foray into broadcasting, and *Slate* magazine, edited by former *New Republic* editor Michael Kinsley, is Microsoft's most well known electronic publishing effort.

Kinsley presumably brings to his job at *Slate* a sense of the journalistic standards that every professionally astute editor in print journalism must uphold. These standards have been essential to ensure quality of information in democratic societies, which are highly dependent on fair and objective reporting for the preservation of an informed citizenry. But as more and more computer executives, in an unholy alliance with the entertainment industry, begin influencing content creation in New Media ventures, what will happen to these traditional standards? And, furthermore, how is the so-called merging of news and entertainment affected (or more importantly not affected) by this trend?

In terms of New Media, nowhere is the flagrant disregard for norms and strictures seen more clearly than in the pages of *Wired*. Interestingly enough, the magazine, whether because of its self-styled irreverence or for some other obscure reason, manages to get away with journalistic transgressions that would be thoroughly castigated if the offender were the *New York Times* or the *Washington Post*. But the magazine's upscale and trendy aura manages to convince otherwise clear-thinking observers that such approaches are somehow acceptable.

The most salient examples are the numerous puff pieces that the magazine has published, little masterpieces of blatant self-aggrandizement. These articles have been used not only to promote and sustain the values of digital culture but also to enhance the prospects of the magazine's relatively small inner circle of writers, contributors, industry insiders, and hangers-on. For example, the magazine produced a cover story about its own investor and senior columnist Nicholas Negroponte for its November 1995 issue. In another case of back-scratching, the magazine published an article on Stewart Brand's business venture, the Global Business Network.

Perhaps the most egregious example of *Wired*'s infatuation with itself and its own narrow community of interest was a June 1994 piece by Joshua Quittner. The article was, as we have seen, a fawning, over-the-top apotheosis of the EFF board members, many of whom were associated with *Wired* in some capacity. The EFF board members at the time included Esther Dyson, Stewart Brand, John Barlow, and John Gilmore. Quittner's article began with a vivid firsthand account of a board meeting that had taken place at a posh San Francisco restaurant. It went on to discuss how the EFF not only had significant designs on shaping cyberspace legislation but was even thinking about transforming itself into a political party. Quittner then traced the roots of the specious connection between digital culture and the sixties by making an almost laughable comparison between this elite group of industry insiders and the legendary Merry Pranksters of sixties fame:

> Who are these guys? In some ways, they are the Merry Pranksters, those apostles of LSD who tripped through the sixties, in a psychedelic bus named Further, led by Novelist Ken Kesey, and chronicled by Tom Wolfe in the *Electric Kool-Aid Acid Test*. Older and wiser now, they're on the road again, without the bus and the acid, but dispensing many similar sounding bromides. Turn on, jack in, get connected. Feed your head with

bits pulsing across the cosmos, and learn something about who you are. . . . Among them are former acid-heads turned millionaires: ideologues who came of age during the 1960s, then proved themselves in the marketplace, Ross Perot–like. And now, grown up, they have retired from the business world to make federal policy, to change the things they were powerless to change when they were love-bead-festooned kids.[6]

Stewart Brand was indeed one of the original Merry Pranksters, but nevertheless the connection is flimsy and cosmetic. For one thing, the Merry Pranksters had been more of a media event than a group involved in serious social change.

In the article the author and *Wired* are milking the sixties connection for all that it is worth in spite of stunning inconsistencies and a blatant values disconnect. In a nicely crafted Orwellian inversion, Quittner works the reader over by attempting to sell the notion that the members of the EFF board, most of them wealthy, powerful, and influential, are somehow comparable to hippies. The notion that they have "proved themselves in the marketplace" subtly suggests that their reborn selves are superior to their implicitly pathetic and "powerless" sixties selves. The article attempts to portray a group of consummate industry insiders as outsiders by transparently invoking a few shopworn sixties artifacts. There are myriad other examples of this phenomenon for anyone interested in unearthing them. Suffice it to say that *Wired*'s appetite for self-promotion was huge, and its journalistic ethics in pursuit of this goal were often questionable.

From a print media standpoint, another interesting twist related to the fact that in *Wired*'s narrowly bounded world the people who were making the news often ended up reporting on it. Thus, Esther Dyson the newsmaker and adviser to Vice-President Gore on communications policy interviewed Newt Gingrich for the magazine's cover story after undergoing a makeover into Esther Dyson the reporter. This kind of blurring of journalistic boundaries is a common occurrence in the magazine's editorial repertoire. The cozy confluence between observer and observed—coupled with a reckless disregard for facts, a penchant for impulsive

impressionism, and a disdain for the traditional checks and balances of journalistic practice—results in editorial content that often falls short of being
balanced, fair, and accurate.

We might speculate that this kind of editorial myopia was abetted
by an ingrained penchant for elitism, coupled with the editorial staff's
naiveté in matters of traditional journalism. More likely than not, this
myopia also related to a misguided sense that the New Media were in some
way superior to the print and broadcast media that they were intent on
replacing. This being the case (the logic would go) it was in some sense cool
or cutting-edge to break the traditional rules of the print media, which were
dinosaurs-in-death-throes and simply could not compete with the overriding presence and significance of the New Media.

Wired and digital culture justify their disdain for traditional media
by declaring that they are owned and controlled by large media empires
and are based on the outdated one-to-many broadcast model instead of the
Internet's many-to-many model. In one issue, for example, *Wired* published a piece by Jon Katz—one of the most influential media critics in
the country—entitled "Online or Not: Newspapers Suck."

In digital culture, information from the broadcast media is generally
held suspect because it is filtered by editors and other information managers rather than pure, undiluted, and aboriginal, as it is on the Internet.
That information on the Net is often spurious, disingenuously framed,
deliberately falsified, or spin-doctored does not seem to be taken into
account in this mode of thinking (see The Drudge Report). But it is this
disdain for the media that forms the basis for *Wired*'s supposedly "irreverent" attitude toward power elites. For some reason, however, *Wired* does
not extend that same attitude to the power elites in the new world of computers and communications.

The Internet
and Spirituality:
A Strange Brew Indeed

One of the oddest notions in digital culture is the vague claim that the Internet and spirituality are somehow linked, an idea reinforced periodically in *Wired* and *Mondo 2000*. If we were to assume, as *Wired* contributing editor Jon Katz once observed, that science has indeed become our national religion, then we might begin to view digital culture as a kind of cult expression of that religion. If this is the case, then we are dealing with a confusing phenomenon, one that is difficult indeed to deconstruct.

I would suggest that not only are many of the values of digital culture not "religious" but they might well be described as ethically challenged. A basic analysis of the content of both *Wired* and *Mondo 2000* suggests that many fundamental ethical notions are undermined by dehumanizing technoid fantasies, such as cybersex. And while the technologies about which these articles speculate, things like microchip implants and cryonics, are not necessarily patently unethical on an individualized basis, they may indeed be so when considered in a larger societal sense.

Langdon Winner, a respected columnist for MIT's *Technology Review*, has questioned the "blasé amorality" of both *Wired* and *Mondo 2000*: "What makes these magazines at once fascinating and appalling is their blasé amorality. One might suppose that the arrival of these powerful technologies for transforming human experience would be an occasion to ponder serious choices and select fruitful possibilities, as distinct from hideous degradations. But in today's cybercult thinking, outcomes are announced, not debated."

There is indeed an almost demonic, certainly grotesque quality to many of the futuristic fantasies of fringe digital culture that flies in the face of anything we could describe as a noble human impulse. The range of cybersexual phantasm found in magazines like *Future Sex* takes pornography to the next level and ennobles it under the pretense that it has metaphysical value. Machine worship takes on a dark, sexual undertone that meets off-kilter occult spirituality on a continuum. And materialism itself implodes and pulls the spiritual nature of human sexuality into the black hole of its own self-obsession.

The notion of cybersex put forth by commentators such as Howard Rheingold and accepted in cyberculture suggests a meandering into dangerous ethical waters. This notion, and others like it, are being peddled under the rubric of bold, anything-goes futurism. In fact, these images and values flunk a litmus test for basic humanism and suggest linkage between certain types of extreme cybercult thinking and the raw edges of social pathology.

Donning virtual reality suits and having sex may sound cool to the readers of *Mondo 2000* or *Wired*, but when the merest bit of evaluative distance is applied, this practice looks altogether too close to the dehumanization found in early science fiction and the works of sociologists who thought about such things back in the late fifties and early sixties. A much truer exercise in values clarification is seen not in contemporary science-fiction writers who shill for a dark social entropy based on sustained overdoses of the technological sublime but rather in the works of writers like Ray Bradbury, who, in books like *Fahrenheit 451*, looked at the dark side of our emerging technological prowess.

When the sex act is reduced to a depersonalized technocratic transaction comparable to a visit to an ATM machine, something has gone terribly wrong with our thinking. Simply by contemplating these bizarre scenarios we have invited dehumanization into the living room. When we look at the depersonalization of sex in light of recent societal trends toward the secularization and technicization of reproductive processes and the increasing commodification of human relationships, the scenarios that emerge are sometimes chillingly Skinneresque. In *Walden Two*, behaviorist B. F. Skinner depicted a hyperefficient society that had made raising children a scientifically controlled collectivist act and state function. This achievement was science's ultimate triumph over government, biology, and everyday life.

The telltale indicator in cybercult thinking that it involves tacit acceptance of a fully secularized technocratic society is the reduction of transcendent and essential human expressions of spirit and relationship to mundane marketplace transactions—to "I-it" as opposed to "I-Thou" relationships, in philosopher Martin Buber's terms. Traditional humanistic and religious concerns about the inherent sanctity of human life are difficult to find in this mind-set. In fact, a strange neo-Gnostic disdain for the human body is evident, and this disdain, based on a desire to be "more than human," is just a few philosophical justifications away from disdain for humanity itself. This represents an impulse to move vertically away from the muck and mire of human origin based on female sexuality and to ascend into an imagined virtual purity bolstered by the power of technology.

Disdain for the human body is premised on the pervasive notion in digital culture that the new computational biology being developed by people like Danny Hillis will eventually result in a further evolutionary step in which human beings—with their imperfect, biologically vulnerable bodies—will be replaced by silicon-based life forms. In this way of thinking, this replacement is the next step of evolution and is vastly superior to human biology with all of its perishable, transient qualities. (In their self-congratulatory hubris, it apparently has not occurred to these theorists that evolution will probably proceed quite nicely without their planning or promoting it.)

This "disgust" with human biology, as Mark Dery describes it in *Escape Velocity*, is common enough in digital culture, but it is also one of the less desirable aspects of many Western organized religions. Thus, cyberculture can be viewed as a mock religion that shares many of the qualities, vertical structure and the search for an ascent experience, common in Western religions. There is no rhizomatic Zen lateralization here, only yet another grand quest for the transcendent experience—not by way of avatars, angels, and other spiritual guides but through machines become godlike with an intelligence that surpasses our own.

There is a quest for both power and perfection in all of this. And there is little in common with true spiritual practice that seeks to lose, not enhance, the ego. Whereas Zen Buddhism warns the initiate about the use of superpowers on the quest for enlightenment, extreme cybercult thinking—with its roots in Nietzschean philosophy, as Dery points out—directly or indirectly seeks this transformation. In doing so, this thinking creates yet another digital mythology, the flight from the body into the virtual. This is the same flight that we see in cyberspace addiction, power cocooning, and in the mythic and thematic structures of latter-day cyber-science-fiction writers like Bruce Sterling and William Gibson.

Much of this muddled thinking about religion comes from digital culture's obvious misunderstanding of the patterns of exploration, initiation, and validation found in archetypal (as opposed to contemporaneously manufactured) myth and the spiritual quest itself. Even Dery, who frequently frowns on the goofier excesses of cyberculture, betrays his ignorance of these traditions when he states "in *The Future of the Body,* Michael Murphy argues that 'metanormal abilities' can be cultivated through a host of 'transformative practices.'"[7] Here Dery fails to understand that these powers were not the end point of the spiritual journey but only a rather useless by-product that must eventually be surrendered.

The disdain for the merely biological (hence the desire to develop the neobiological) is easily seen in the frequent use of the word "meat" to refer to the human body. Both Rheingold and Barlow use this term, and Barlow uses the term "meat space" to refer reductionistically to the physical world, which he contrasts with the purer world of cyberspace—a world uncontaminated by the physical body and its imperfections. Both Barlow and Kelly continually conflate the pseudoetheric realm of cyberspace with the realm of true spirituality. No longer a sacred temple, as described by many religions, the body is now seen by digital culture as suffering from planned obsolescence, as a useless appendage or a failed piece of engineering developed by a discredited creator. Like a worn-out automobile, it will be replaced with something newer and shinier.

This sadly reductionist view of the world is fully consistent with scientific materialism's eagerness to play God via gene tinkering and the patenting of human genetic material. Digital culture appears to be very comfortable with these trends, and sophisticated techniques for bioengineered systems and other emerging hyper-technologies are in fact held out as the only way to stem the tide of environmental degradation, end world hunger (something that science promised was going to happen thirty years ago), and provide a technological quick fix for many otherwise intractable social problems.

Are there historical archetypes for what proponents of digital culture and scientists like Marvin Minsky and Hans Moravec are trying to accomplish? It would seem so, but we can find these archetypes only by reaching into the increasingly side-railed mythical structures of history. The attempt to raid the heavens is a consistent and recurrent theme throughout Western literature and mythic tradition, from Goethe to Milton to the Bible to Greek mythology. In more-contemporary literature, there are similar themes. Some of them have shifted into the realm of science fiction, where, as William Irwin Thompson points out, mythic narratives are often reformulated for modern tastes. The mad-scientist theme is the most obvious example. In this theme, the scientist longingly covets divine power over life and death and decides to don the mantle of creator, as in the case of Mary Shelley's classic, *Frankenstein*.

Promethean overreach and human hubris in the pursuit of divine knowledge are, in fact, two of the deepest and most enduring archetypes in our mythic past. It is fascinating, then, to consider that at the cusp of the new millennium, some are putting forth serious arguments in favor of rekindling this raid on ultimate knowledge. Stewart Brand's "we are as gods and might as well get good at it" speaks volumes in this regard, as does a chapter in Kevin Kelly's *Out of Control* called "The Nine Laws of God." Kelly has commented elsewhere that "To be a god, at least a creative one, one must relinquish control." However, this new and secularized attempt to create pseudospiritual and humanistic structures out of the dying embers of scientific materialism may encounter, in the Hegelian sense, a powerful counterforce—a resurgence of nativistic spirituality and humanism. Even if it is ignored by contemporary media, this clash is the basis for Sven Birkerts's observation in *The Gutenberg Elegies* that the central argument of our time is "between technology and soul."

The EFF

and Net Politics:

Technocracy in the Making?

In the realm of Net politics, one of the most interesting (and potentially disturbing) notions is that the Internet might actually serve as the basis for wholly new types of political parties. In any number of venues, digital culture has been talking about this concept for some time. What is interesting in such considerations is the potential for a unique fusion of value systems, the merging of technology, culture, and politics. Such a fusion, of course, represents the very essence of technocracy.

As we have seen, the EFF was reportedly considering the option of forming a political party a number of years ago. This possibility brings up an obvious question: How could a group with such a narrow constituency presume to think of itself as capable of representing the broad spectrum of interests that go into any meaningful definition of a political party? Deconstruction of this mind-set holds the key to understanding what this apparent movement toward technocracy is really all about.

Net politics is not simply a figment of digital culture's imagination; it has been discussed in serious policy circles in Washington and elsewhere. A 1995 Department of Defense study authored by Charles Swett took a hard look at the domestic and international implications of the Internet and other expressions of the electronic polity for American democracy. The report, entitled *Strategic Assessment: The Internet,* offered the following observation: "The political process is moving onto the Internet. Both within the United States and internationally, individuals, interest groups, and even nations are using the Internet to find each other, discuss the issues, and further their political goals."[8]

If true, this is a rather stunning development in the history of the American political system. Embedded in these observations are two interesting and—from the point of view of conventional politics—rather radical notions. The first is that, in ways yet to be defined, many of the functions and institutions of government could be replaced by the Internet. (Perhaps it would be more accurate to suggest not that the Internet is replacing the political system but rather that the system itself is being virtualized, that is, moved into the venue that is the electronic polity.)

There are many areas of government activity where this shift might make sense, if properly implemented. A relatively large number of routine and time-consuming processes could indeed be automated via computer technology. Such automation could save taxpayers time and money by eliminating costly and unresponsively bureaucratic transactions and providing more-user-friendly government "service" to citizens. For example, driver's license or registration renewal could be accomplished electronically instead of by means of the usual dreaded trek into the stark interior and endless long lines of the local division of motor vehicles. Those without computers at home could instead access information kiosks at designated public areas such as libraries or post offices.

The second notion that arises from the assumptions of the Defense Department report is, however, far more problematic and complex. In the new digital politics, the Internet becomes, in essence, a substitute for the representative form of democracy we now have in place. This substitution would presumably happen under the guise of electronically implemented, direct, participatory democracy.

To implement this change, Alvin Toffler and Newt Gingrich—two advocates of digital democracy—have considered restructuring the Constitution in order to bring it in line with Toffler's "Third Wave" (that is, information age) perspective. This restructuring suggests radical, sweeping, and fundamental changes to the American political system based on the untested and untried capabilities of the Internet as a mechanism for citizen feedback.

What is especially troubling is that the new technocrats who have been arguing in favor of these changes, with characteristic missionary zeal, want to push these ideas forward whether or not the American populace is ready for them or even fully up to speed on their significance. I suggest that digital politics, while not the backdoor politics of smoky rooms and slush funds, is a new kind of social engineering that is every bit as hidden from the public view and equally unresponsive to due process.

Although the ideas of the new technocrats have yet to reach a wide audience, they can be viewed as the first glimpses of a trend toward technocratic governance, and they are being considered in places where real political power is leveraged—for example, in a 1995 conference on the future of technology. This high-profile inside-the-beltway event involved a number of the usual suspects: Gingrich, Toffler, the Progress and Freedom Foundation (PFF), and the EFF. According to technology writer Mark Stahlman, the EFF and PFF have discussed the implementation of electronic democracy a number of times over the least several years, and the 1995 PFF-sponsored conference was one of the crowning moments of collaboration between the two organizations. The conference, televised on C-SPAN, included the conspicuous participation of the two EFF cochairs, Esther Dyson and John Barlow. During the meeting, Barlow, ever the semanticist, repeatedly used the phrase "terrestrial politics," presumably as a means of distinguishing between conventional government and the new, distributed, decentralized Net government envisioned by cyberlibertarians.

The PFF meeting had an odd, radical disjointedness to it. It was interesting to see whom the principals of Gingrich's favorite think tank had invited to put forth their views on the future of computer networking and telecommunications. In addition to Dyson and Barlow, Stewart Brand and Kevin Kelly were also among the participants. Also present was a group of industry executives that included General Magic's CEO Marc Porat and Bell Labs's Arno Penzias.

Once again, the relatively monolithic character of digital culture was clearly evident from the proceedings, since there were few in that group, with the exception of a single congressional representative, who put forth any serious divergent viewpoints in favor of "terrestrial government." The group was clearly the "down with government, up with networks" crowd.

What was interesting to observe in this new confluence of technology and politics was how new forms of political power were already being generated from these two volatile fuel sources. Especially troubling is Barlow's facile and logically inconsistent mixing of politics, technology, and a kind of homegrown metaphysics, a mixture that was perhaps a little too high-octane to be aired in the political realm.

Barlow likes pushing his patented brand of Net metaphysics into realms where it really does not belong. Kelly's "hippie mystic" is in reality a longtime Republican and a right-leaning radical who excels at laying down the "philosophical" justification for replacing conventional politics with Net politics. Like some of the other thought leaders of digital culture, Barlow enjoys nominal counterculture credentials as a former lyricist for the Grateful Dead and, like Stewart Brand, milks that connection for all it is worth.

Beyond this camouflage, however, lies a rather unflattering portrait, at least according to some observers. Writing in the November 1995 issue of *Upside* magazine, G. Pascal Zachary says, "Of all the irrationalists attracted to the Cyber-Church, John Perry Barlow is the most amusing. He is the most audacious, the most crass; he combines the zeal of Johnny Appleseed with the intolerance of a Nazi." Zachary goes on to suggest that Barlow's stance is part posturing, part business opportunity in that he "makes a good living from telling people that even he can't make sense out of the digital sea change."

Where exactly the EFF stood on network politics was difficult to say. Nominally it was a public interest organization working to shape enlightened cyberspace policy and legislation in Washington, but the group went through some turbulent readjustments and eventually came to be characterized by its critics as just another inside-the-beltway lobbying organization.

The willingness of the lamb to lie down with the lion became evident. Although the EFF shared many policy platform issues with the Computer Professionals for Social Responsibility (CPSR), many within the CPSR would eventually come to look skeptically at the EFF's activities. Unlike other public interest groups that sought to remain independent from undue outside influence, the EFF seemed to have no qualms about receiving its funding from large industry players, including the powerful regional Bell companies.

If the EFF/*Wired* mind-set is either a crucible for or an early manifestation of a new technocratic elite, then there is another bothersome element in the mix: the fact that these groups appear to be operating in a kind of virtual collusion that is only rendered into the art of the possible by the technology we now have, let alone some future version yet to be announced and deployed. While the EFF is a nonprofit organization, it operates largely in the online environment, where opinions are reinforced and allegedly grassroots support is marshaled. This practice, in itself, seems exclusionary for a public interest group.

The EFF may be a model for a new type of organization that will have roots in both the real world and the virtual world. Yet if it is a model for the future, it raises serious questions about accountability and access. When I was a member of the press, I found traditional access to the EFF to be difficult. Because of its self-selected modes of communications—distribution of position papers and articles via e-mail and conferences on the WELL and other systems—it is elitist. Much of the EFF's influence is due to its online pamphleteering and opinion shaping. Only those who have access to these electronic systems and services and, more significantly, travel in the right circles of cyberspace are in the mainstream information flow of this organization. The disconnected connectivity of online response is clearly the EFF's preferred modus operandi. The organization deals with channels of information it chooses to utilize, rather than making more-conventional modes available to the general public.

Is this how lobbying groups will operate in the future? Is it too much of a stretch to wonder if an online venue could become the new smoke-filled back room of the Washington political scene? Clearly, the notion of a technocratic elite is not compatible with the idea that electronic democracy will be more democratic than the system we now have. The fervor of the digital elite for this notion should also make us skeptical—electronic democracy easily has the power to become crypto-exclusionary. Some television stations are already taking Web polls on serious political issues, and this kind of polling de facto excludes lower- and middle-class Americans. An increase in the number of polls of this nature would be a disturbing but not an unlikely trend in the current environment.

Telecom Unchained:

Privatizing

the Public Network

The emphasis on freedom from government interference in cyberspace found in *Wired* magazine and other vehicles of digital culture is wholly consistent with a tendency for the computer community to embrace libertarianism. Libertarianism comes in various flavors, but the kind associated with *Wired* is characterized by right-leaning positions on many social issues and a strong emphasis on deregulation, free markets, and keeping the government out of the activities of ordinary citizens. Paulina Borsook describes the phenomenon nicely: "The convergence between libertarianism and high-tech has created the true revenge of the nerds: Those whose greatest strengths have not been the comprehension of social systems, appreciation of the humanities, or an acquaintance with history, politics, and economics have started shaping public policy. Armed with new money and new celebrity—juice—they can wreck vengeance on those by whom they have felt diminished."[9]

The extent to which techno-libertarianism is also consistent with the new utilitarianism is not surprising. However, arguments that the government should back away from the expansion of the Internet are not without merit. Overlaying the growth of the Net is an increasingly outmoded and complex set of government regulations that have, like layers of sedimentation in the Grand Canyon, been built up over a long period of time. The purpose of these regulations was originally to watchdog the telecommunications industry when it was primarily a voice-only network operated by what was then the world's largest corporation—AT&T. Online services, data networks for corporate use, and the Internet and World Wide Web as a general purpose network are, however, in the long history of telecommunications, relatively new phenomena.

In general, as telecommunications became more complex and important to the nation's economy, it also attracted the involvement of many state and federal regulatory agencies: at the state level, public utility commissions, and at the federal level, the Federal Communications Commission (FCC), the National Telecommunications and Information Administration (NTIA), Congress and the House Subcommittee on Telecommunications and Finance, the Justice Department, and the courts.

These complex and sometimes duplicative regulations, coupled with an increasingly outmoded regulatory model for policy making, increase the chances that government efforts will slow down or even quash the natural growth of the Internet. The other danger for the development of the Net is that many regulators appear not to understand its labyrinthine complexity, preferring instead to view it through the old lens of a regulatory structure put in place to fulfill an entirely different set of objectives.

In this sense, the libertarian mind-set has some legitimate concerns about government being too heavy-handed with respect to the Net. But digital culture also believed that the free market could do no wrong and that there was little need for public policy in the new world of telecom unchained. This reflected a fundamental misunderstanding of the true nature of large-scale, complex systems such as the communications infrastructure; and it patently ignored the cooperative basis upon which these systems were built by means of standards and other cross-industry mechanisms to ensure interoperability of the many technical systems involved.

Digital culture also abhorred the notion of government subsidies and, for that matter, any scheme involving the redistribution of wealth. Instead of focusing on a constructive attempt to establish new policies and structures for the emerging Internet, as the CPSR had done, digital culture opted for a kind of barely disguised social Darwinism, predicated on the law of free markets and the whims of corporate interest. With its own access to the Net securely in place, digital culture signaled its contentment with the status quo and its desire to allow the Net to evolve in its own fashion, without policy, purpose, or planning.

This rather naive system of belief was based on pure supposition: the supposition that a chain reaction of events would bring about some sort of techno-utopian fulfillment. Free markets would foster competition, competition would foster lower prices, and lower prices would ensure that everyone would eventually have reasonably priced access to the Internet.

Since belief was enough, action was unnecessary: All of this would occur without the assistance of public policy groups, opinion shapers, policymakers, or legislators. The entire lot of policy wonks could vanish into the Bermuda triangle, and we would all be better off without them. Besides, all that thinking just gets in the way of progress.

Unfortunately, this exuberant but ill-founded techno-optimism failed to take into account a basic reality: There were huge corporations within the telecommunications industry that would eventually get into the Internet business and gain considerable influence over the Net's destiny as a means of social communication. Who would provide the counterbalance if corporate excesses skewed the Net in the wrong direction? This optimism also ignored the fact that in the new environment wrought by the Telecommunications Act of 1996, there was good reason for concern about whether real competition was even going to happen, let alone whether bandwidth would become essentially a free commodity, as in the cornucopian visions offered by George Gilder.

In general, *Wired*'s view of how the Internet should evolve and how the future of the telecommunications industry should take shape lined up remarkably well with the thinking adopted by Third Wave industry conceptualizers like Newt Gingrich, Alvin Toffler, Peter Huber, Gilder, and other shoot-from-the-hip policymakers whose political leanings could be described, depending on the issue, as right-leaning or quasi-libertarian. This alignment is yet another chink in the armor of the contention that *Wired* is carrying on the legacy of the sixties.

The belief systems shared by these individuals were every bit as techno-utopian as *Wired*'s, although they were delivered with more sophistication and panache and the seasoning of experience. All strongly believed that, in some way that could not quite be explained, the new Internet-based economy was single-handedly going to save society from its most abiding social problems. In reference to Gingrich and his shared affinity with digital culture, *Time* magazine quipped: "The merry cybernaut wants a new moral order, not just a new political one, in which the poor will find their salvation on the Internet, and private charities will succeed where government bureaucracies have failed."[10]

Of deepest concern in this kind of thinking is the manifest tendency toward privatizing the very public function that is telecommunications. Such concerns are not by any means abstract: They are already being widely implemented, and digital culture has done its part by providing the necessary ideological support for their implementation.

As the electronic polity develops, many social functions will become highly dependent on access to networks, Internet-based or otherwise. Deregulation, accomplished in the proper manner, is indeed long overdue and has the potential to bring about more competition in the rapidly expanding new communications environment. But deregulation and free market ideology alone are not the answer. It is already a matter of common concern that the telecommunications reform legislation passed by Congress has not implemented deregulation in the right way.

The danger is that this kind of thinking will, in the long run, create not communications platforms that are robust and have open architecture but rather the electronic equivalent of gated communities: virtual and special interest communities that will further partition society and—perhaps most insidious of all—do so in a manner that will be invisible to those who are not "well connected."

An even more important concern, one that is hugely underestimated, is that from a pragmatic and administrative standpoint, a privatized communications environment might actually function far less effectively than the public telephone system that has served us so well for so many years. We run the risk of developing islands of incompatible systems of communication that will have grave impacts not only on the prospects for electronic democracy but on democracy itself.

While competition is a superb value in our capitalistic system, we must have a clear perspective on the corresponding need for cooperation in the development of the technology systems on which we all depend. This notion of cooperation, absent from the lexicon of digital culture today, must become a watchword in the next millennium. Without cooperation, the great technological accomplishments that we have built and seek to build run the risk of collapsing in chaos and the fragmentation of self-imposed isolation.

Science, Culture, and the Internet

The resurgence of religio-politics around the world may seem to have little to do with the rise of the computer and the new economy. But it does.

—Alvin Toffler

Spiritual nourishment is more important than mere industrial productivity.

—Lewis Mumford

The impersonal will become personal.

—Nicholas Negroponte

I buy the magazine [*Wired*] and read and study it in order to engage myself in what I think is the central argument of our time—the argument between technology and soul.

—Sven Birkerts

Is Science

Our National

Religion?

In the March 1995 issue of *Wired,* media critic Jon Katz asked: "Is technology a good witch or a bad witch? . . . In this country, where faith in technology is the closest thing we have to a national religion, and in the new media culture where belief in technology is a religion, it's a riveting question." Katz was emphatic and unblinking in his answer. There seems to be little doubt in his mind that science has become, in the words of cultural historian Morris Berman, the "ultimate arbiter of reality," a belief system so powerful and pervasive in the cultural lexicon and popular imagination that it competes on the same existential turf as, and therefore meets the technical definition of, religion.

I have argued elsewhere that it is in the values systems of digital culture that a new kind of technocratic direction for society is being incubated. The idea of science-as-religion is important in this context because our underlying cultural values have the potential to foster such a change, and therefore we need to peer behind the curtains and ferret out the invisible ropes and wires that hold up our more cherished assumptions.

As a technologist, albeit one with humanistic leanings, I have tried to make the case that we should be free to use specific technologies as tools without adopting technology as lifestyle. I make these arguments as someone who wishes to use the Web and the Internet as tools but on my own terms, as someone who is not interested in participating in the recycled scientific materialism that underlies much of digital culture. (Nor do I wish to worship in what G. Pascal Zachary calls the Cyber-Church!)

We must, however, ask whether it is possible to be that selective? How much linkage exists between the culture of technology and the technologies themselves? Are there real and specific choices to be made in adopting the new modalities, choices that will allow us to express individual values and creativity—or are we simply buying into a system of thought and culture that will, as Stewart Brand suggests, steamroll everything that it does not deem to be a part of its own mega-regime?

The question of whether science has become our national religion is closely related to the issue of technocracy, but a few distinctions are in order. The first issue is existential, that is, a matter of root values, whereas the second is social, political, and cultural. To some extent, the adoption of science as the paramount system of beliefs (one that can seemingly coexist happily yet paradoxically with a religious system of beliefs) is a precondition for a technocratic system of governance.

In *Technopoly,* Neil Postman argues that the United States has already moved well into a technocratic mode of governance. He categorizes societies as either tool-based, technocratic, or technopolistic and posits that "the United States is the only culture to have become a technopoly." Although Postman's distinctions seem a bit artificial and mostly a matter of degree, they are a useful starting point. That said, I do not agree with his contention that the United States is a technopoly—the most extreme form of technocratic governance. We are, however, rapidly heading in that direction, and it is not too hard to spot the directional arrows. Looking at areas such as health care and education, for example, we can pick up the trail.

Rapid and poorly managed technicization, in tandem with other factors, has already wreaked havoc on our health-care system. Our educational system is in chaos, and as Camille Paglia has pointed out, the discussion of education has been largely stripped of humanistic and "big picture" considerations in favor of technology-du-jour issues. Practically every speech about education, whether made by governor, legislator, or educator, stresses technology as the bottom-line solution.

The degree to which we as individuals feel that technocracy is viable or imminent is largely a function of the degree to which we perceive technology to be a dominant force in everyday life. Personal vantage point is critical. The wage earner who tends lawns for a living or the fast-food service worker will have little visibility into this issue. To many, technology *is* invisible, a background environment in which microwaves hum, blenders blend, computers connect, and other barely noticed microevents take place by the hundreds and thousands every day. In this mode of perception, technology has been blended seamlessly into the fabric of everyday life and is noticed only when one of its more obvious or essential functions fail. Arguably, much of our technology has this quality of invisibility. It cannot be deemed intrusive because it does not seem to intrude at all; instead it is the tireless, invisible servant of humankind.

Of course, the playful logician can just as easily arrive at the flip side of this perception: The fact that technology has woven itself so deeply into our lives that we barely notice it suggests that it has indeed conquered culture. Its invisibility is proof that we take it for granted as an embedded aspect of contemporary life.

Individual perceptions aside, I would argue that there is no question that science and technology have attained a curious primacy over culture in the United States. This development has some normative and healthy aspects; but it is also cause for concern. In particular, technology's more utilitarian qualities have become part of what it means to be an American. But as technologically driven as our society already is, what is important is not just the degree to which technology has physically permeated our lives but also the deeply embedded set of assumptions that we, as a nation, have internalized.

It seems clear that we have internalized the belief that science and technology are primarily responsible for social progress and improvements in quality of life. We have done so without regard for the complexities of testing this thesis against hard reality. Coupled with the classically linear Western view of history as social progress over time, our notion of well-being is intertwined with our belief in a beneficent science that brings health and economic plenitude into our lives. So powerful is this belief— and here the belief system mirrors the "irrational" dimensions of religious thinking—that we effortlessly and with a facile but well-practiced casualness exclude the many failures of science and technology from a true reckoning of their worth.

If technocratic governance is our destiny, it will be brought about by a confluence of several powerful trends; the strength of the technological diaspora alone is not, in my opinion, sufficient to cause it to happen. Certainly, the advent of the Internet and other new technological regimes has brought the realm of science and technology to our doorstep. We are poised to foster these technocratic tendencies unless we as a society can restore a sense of balance. We must strongly and consciously counterbalance every step we take toward technological enhancement with a grounding in the real and the physical. The virtual can allow us to soar, but only if we maintain our ties to a natural and biologically derived sense of self.

Back to the Future:

Science Fiction

as Mythology

In *Powershift,* Alvin Toffler describes the central global cultural conflict of the current age as a tug-of-war between two forces: secularism and religion. Although Americans have always been guaranteed freedom of religious expression, an uneasy tension between these two forces has always existed in this country. The volatility of issues such as school prayer and abortion offer adequate testimony to this fact. Interestingly, as we head into the new millennium, both forces seem to be gathering momentum, as if heading for some final showdown.

Toffler also points out that secularism is an essential ingredient of modern democracy, and certainly the same could be said for technocracy. Technocracy represents a fusing of technology and politics. But it also can encompass strong cultural components as well. As we have seen, the American sense of well-being has long been bolstered by scientific idealism and the pursuit of secular perfection. "Progress is our most important product," Ronald Reagan frequently assured Americans while he was a spokesman for General Electric. That message contained several seeds of cultural myth: a reaffirmation of belief in the continuously improving future and the notion that this future was highly dependent on technological growth.

Every culture is shaped by mythologies, whether religious or secular. However, as Sharona Ben-Tov points out in *The Artificial Paradise,* the United States may be unique insofar as its mythic structures are not only highly secular but also oriented less toward the past than toward the future. Given this context, we can see the role of digital culture far more clearly. If the techno-utopianism of digital culture is viewed as a kind of laboratory for forging new digital mythologies, then its role as a cultural reinforcer becomes evident.

Science fiction plays a strong role in the cultural dynamic cited by Ben-Tov. There is also an interesting and strong connection between the techno-utopianism of digital culture and contemporary science fiction. If American cultural mythology is oriented toward the future, then it is science fiction that represents an important creative wellspring for those mythologies. (Witness, for example, the immense popularity of *Star Trek* and *Star Wars* in the cultural landscape.)

Ben-Tov reinforces this connection in *The Artificial Paradise* using the classical tools of literary criticism. By looking at digital culture in this light, we see it not as a new phenomenon but as the culmination of a strain of thinking that has been a part of the American experience for a long time.

Among the modern science-fiction works that have most influenced digital culture are *Mirrorshades* by Bruce Sterling and *Neuromancer* by William Gibson, from which the term "cyberspace" was derived. This dystopian science fiction, however, contrasts sharply with the science fiction of earlier decades. Conspicuously absent in these modern works is a sense of philosophical exploration of the social consequences of life in various imaginatively presented technological regimes. Gone is the querulous uneasiness with aspects of the future deemed dehumanizing or otherwise undesirable. In its place is a desensitized and almost blasé acceptance of postbiological social decline. There is a survivalist, frontier quality to the science fiction favored by digital culture, and, thematically, survival generally depends on possessing the proper degree of do-it-yourself technological prowess. Unlike the science fiction of the fifties and sixties, modern science fiction renders few implicit or explicit value judgments about the presented realities of alienation and deep societal disconnection.

During the 1950s, the notion of alienation was commonplace in both fiction and nonfiction. For example, consider sociologist Émile Durkheim's notion of "anomie"—the sense of separation from primal experience that was seen as a necessary but unfortunate condition of modernity. Ironically enough, these concerns were being raised when the United States was in some ways the closest it has ever been to actually achieving the ideal of techno-utopianism. The postwar American

standard of living was indeed high, and life in the rapidly expanding sub-
urbs was made increasingly and seductively comfortable as new and affordable consumer products such as dishwashers, air conditioners, washing
machines and dryers, televisions, and automobiles flooded the marketplace. Except for those pesky concerns about nuclear annihilation, life
was good. Given this context, however, it is all the more interesting to
note how concerns about the loneliness and alienation of American society were being raised by thoughtful and respected social commentators of
the period.

The science fiction of the fifties and sixties was thoughtful and intellectually stimulating. It often raised questions about the direction in which
science and technology were taking us. In *Fahrenheit 451,* Ray Bradbury
painted the grim and chilling picture of a future where people would stroll
through public spaces oblivious to their surroundings and languorously
plugged in to an electronic device of some sort. (There were, of course, no
Walkmans at the time.)

Bradbury also foretold, with astonishing accuracy, the advent of the
virtual world. In his vision of the future, television was a central and dominant life experience, complete with wall-sized screens and virtual family
members. The book depicted a shallow and pleasure-obsessed society in
which an individual and collective sense of compassion had atrophied, in
part because deeper thinking and even garden-variety interiority were
deemed to lead to unhappiness. In this "bring-on-the-Prozac" environment, unhappiness itself was, in effect, outlawed. Books and literary works
were contraband because they tended to produce undesirable chain reactions: Reading led to thinking, and thinking—even more dangerously—
could lead to a condition of somber reflection that was indistinguishable
from depression.

Guided by an Orwellian mandate, "don't worry, be happy," the society Bradbury portrayed was pathetic and infantilized by technology, a
society in which interiority and reflection, not to mention intellectual
exploration, were systematically destroyed for the sake of the greater good.
Unfortunately, it is not difficult to see a few disturbing traces of Bradbury's dark vision in our own contemporary lifestyle.

Just as the cultural somnambulism of the fifties gave birth to penetrating insights about our society's values, so to is the gathering momentum of forces of technocracy leading to the necessary and inevitable reappearance of a suppressed cultural and spiritual nativism. Herein lies the struggle between technology and soul alluded to by Sven Birkerts in *The Gutenberg Elegies*.

Americans sense that postmodern society has, to a large extent, lost its moorings and that a deep struggle to reassert basic human values is underway: a moral and spiritual renewal that involves reestablishing the fabric of community, and a reaction against the smothering uniformity of living in the media-driven claustrophobia of a consumer society. Americans are also looking for a way to restore deeper values that are being eroded by secular scientism and the virtual flight from the body and, most unfortunately, from nature itself.

These trends surface in the mainstream media only occasionally. As more and more Americans discover holistic health alternatives, national drugstore chains have begun to carry extensive lines of herbal medications. In fact, herbal medicine, which challenges conventional medical wisdom on many fronts, has become a billion-dollar industry. Spiritual retreats, meditation, and yoga are back in vogue. Bookstores sell and book clubs discuss books on the spiritual quest, many of them bestsellers such as *The Celestine Prophecy* and the *Care of the Soul*. All of these cultural indicators signal a resurgence of values in the face of what former Pennsylvania governor Robert Casey has flatly called our nation's "moral decline."

Science, once revered by all for expanding our awareness of the universe and bringing many important discoveries, has devolved into secular scientism, a set of values unto itself and now even a lifestyle rather than simply a methodology for empirical inquiry. The means have become the end. Science and technology have overstepped their bounds and become self-sustaining value systems that reach into every nook and cranny of everyday life.

The Internet and digital culture have played a role in the transformation of science by breathing new life into the scientific value system and providing it with an upscale style that sweeps through popular culture like a high-fashion model's glittering cape and blithely declares all new scientific initiatives to be part of the show. Thus, new hyper-technologies—biotechnology, for example—are given ground cover as part of an undifferentiated stew and allowed to gain acceptance without any real evaluation. If the computer is involved, then who are we to question its indefatigable forward progress?

The resurgence of forgotten and swept-aside values will balance this scientism and renew emphasis on the human use of technology. This process has already begun, but in the meantime, we must never abandon our technological exploration. In a way that we have yet to fully understand, this exploration underwrites much of our humanity. But we do need to rescind the blank check that science enjoys and restore a larger perspective. Perspective and balance will be essential to return holism to our view of the world, a holism that can reconcile these emergent digital mythologies with the archetypes of the past.

Mediated Society:

The Cybersomething That Lies

beyond Gesellschaft

It is important to look for both positive and negative trends in the various scenarios that pair technology and culture. I would like to render one specific scenario in which the cultural and political reformulations we have considered might provide unique benefits to our overburdened society. The concept of anarchic pluralism for which the Internet is the putative model has some interesting implications with respect to the need to resolve the fundamental conundrum in American politics described by social critic Christopher Lasch:

> We find plenty of evidence to confirm the impression that the modern world faces the future without hope, but we also find another side of the picture, which qualifies that impression and suggests that western civilization may yet generate the moral resources to transcend its present crisis. A pervasive distrust of those in power has made society increasingly difficult to govern, as the governing class repeatedly complains without understanding its own contribution to the difficulty; but this same distrust may furnish the basis of a new capacity for self-government which would end by doing away with the need that gives rise to a governing class in the first place.[1]

This new capacity for self-government seems to reflect the best aspirations of digital culture's Net politics, although Lasch makes this affirmation without regard for technology of any kind. The creation of this new capacity may indeed occur, and I do not by any means discount the positive dimensions of such a change. Whether it takes technology-as-metaphor either as a means of acceptance or as an actual factor in bringing it to reality, let us celebrate the path to finality as we would in any other circumstances.

Perhaps as we plod into the new millennium we will see the "end of politics" and the emergence of a more efficient, less convoluted paradigm of human organization. We can only hope that it will be based on new forms of cooperation that make significant allowances for diversity. But perhaps, in spite of our hopes, the crypto-conformism inherent in Kevin Kelly's concept of "hive mind" will prevail, in which case humankind will learn to function as a single sleek superorganism dwelling virtually in the bright latticework of the electronic polity. Think Singapore.

The Net's lateralism, its lack of hierarchy, is enticing when viewed in the light of Lasch's comment. But while I am intrigued by the Net's power to advance new forms of social and political interaction, I would be far more sanguine about such scenarios if they were happening under social circumstances different from those of today.

Can anyone suggest that it is impossible to correlate ever increasing electronic mediation with the increase in civil disorder and the loss of traditional social relationships? It is entirely possible, of course, that these two trends have no formal relationship. It is also possible that the Net, if indeed it has some cloaked evolutionary purpose, will function as the scrambler of deep structural codes and that the resulting social confusion and disorder will be a useful prelude to some unspecified but soon-to-be-revealed quantum leap in human organization.

It seems clear that if we are to have some hope for moving beyond the usual and expected entropy of political solutions, we must avoid the obvious pitfall of allowing politics to be replaced by technocracy. Nature abhors a vacuum, particularly a political one.

I remain unconvinced that digital culture's enthusiasm for Net politics is motivated by a compelling appreciation of democratic action so much as a desire of those forgotten by power to have their moment in the sun. Nevertheless, I intend to keep an open mind and advise the reader to do the same. These are complicated issues, and we as a society are just beginning to sort through them.

Every society has choices. The Internet can and should remain a vital tool for social transformation but only if it is offered as a free gift and not as a Trojan horse for technocracy. It is indeed possible, even likely, that a major shift in social organization is already taking place, a shift whose outlines are nevertheless barely discernible through the postmodern haze. But if we assume the ascendancy of new technocratic forms of control, which is hardly implausible, what might our society look like if it fell under the spell of the technological sublime and the pseudometaphysics of digital culture?

Answering this question requires, in part, extrapolating from several societal shifts that have taken place over the last several hundred years. One of these shifts is described in classical sociological terms as the transition from community (gemeinschaft) to society (gesellschaft). This transition, of course, has its roots in both industrialization and urbanization. If we accept either Toffler's Third Wave or Kirkpatrick Sales's notion that the advent of the computer represents a second industrial revolution, then we can at least surmise that yet another shift is underway and that it is unmistakably related (once again) to technology.

Beginning with the definitive work by German sociologist Ferdinand Toennies, the literature of sociology offers an interesting glimpse of the process. We can perhaps see into the future by revisiting a bit of our recent past. According to Christopher Lasch, Toennies envisioned a future in which the world would become "one large city . . . a single world republic coextensive with the world market." This vision should have a familiar ring to the contemporary ear. In *True and Only Heaven* Lasch goes on to delineate Toennies's views on the differences between community and society:

> Consider some of the many contrasting typologies that Toennies piled on the basic contrast between community and "society" or contractual "association." Community rested on feeling, association on intellect. Community appealed to the imagination and the emotions, association to calculating self-interest. Community encouraged belief; association skepticism. . . . The community was an extension of the family, whereas "family life was decaying" under the principle of association. People now confronted each other as "strangers." "Custom, habit, and faith" governed community life, "cold reasoning" the life of the modern metropolis. The community was feminine, the metropolis masculine.

The contrast also corresponds to the contrast between youth and old age, or again, between the common people and the educated classes. Metropolitan life gave rise to a type of thinking and action characterized by the separation of means and ends and exemplified, in its prototypical form, by commercial exchange. Under community, on the other hand, means and ends were inseparable.[2]

Durkheim formulated the classical notion of anomie, which later came to address modern rootlessness and acknowledged that an array of trade-offs was necessarily involved: Community life was confining, and urban life offered a liberation of sorts. In other words, alienation has its benefits, anomie its comforts.

It may be argued that Toennies was anticipating what some would now, with characteristic vagueness, call the New World Order, what cultural historian Lewis Mumford has called World Culture. But the more interesting question is whether, from a sociological standpoint, this trend is propelling society at large *beyond gesellschaft* and toward another form of cultural organization, as yet unspecified.

At this juncture, it seems impossible to tell whether the current shift is a merely temporary phenomenon on its way to something that will incorporate a deeper sense of human community or whether it is a process that will greatly magnify the qualities of gesellschaft to a new and less acceptable level from the standpoint of human values. If we assume that disintegration usually precedes reintegration, this is a tough one to call.

To the best of my knowledge, this phenomenon has not been named or appropriately identified. I have spent some time thinking about how to adequately describe it, feeling all the while a certain level of sheepish presumption in doing so, since I am hardly a sociologist. We might call this new form of human organization virtual society. But the term "virtual" is fraught with semantic peril and sows confusion when juxtaposed against the notion of virtual community, as it pertains to the online experience. The term that I have reluctantly arrived at is "mediated society." This term is intended to convey the notion of a society in which the traditional gestures of human interaction are mimicked by electronic mediation, a society in which a shadow dance is conducted at virtual arm's length from deeper forms of intersubjectivity. The virtual world is, after all, a world of sign and symbol in which meaning is supplied by imagination and projection.

Is a "classical" alienation, then, a defining characteristic of mediated society? My own assessment is that it is. Just as the process of urbanization required us to trade intersubjectivity for a new freedom of movement and personal expression, so too will the move toward a mediated society require that we accept alienation as the trade-off for whatever benefits such a society might provide.

Interestingly, the transition toward a mediated society is in some ways simply a linear extension of the progression from community to society. And just as a certain amount of alienation was acceptable in the gradual shift to gesellschaft, so too will alienation become manifest in the current paradigm shift. The question is: What level of alienation is acceptable?

How can we envision the endpoint of this transition? Imagine gesellschaft, with its alienation and exhilaration, but imagine it not limited simply to the context of the urban environment but extended throughout all levels of our national culture and, indeed, throughout the globe. In the best-case scenario, humankind will "come of age" and the distinctions between haute couture in the urban environment and the sense of deep community and the folkways of rural life will disappear. Perhaps the breakdown of these distinctions will eventually lead to the kind of collective mind meld envisioned by Arthur C. Clarke in *Childhood's End*. In the worst-case scenario, we return to the mercilessly efficient "hive mind."

Whether from the lofty position of a Toynbee evaluating vast civilizational narratives or from the standpoint of individuals caught in the crosshairs of contemporary life, we must ponder the question of whether mediated society has improved quality of life in any real way, whether it has reaffirmed the lasting set of core values that Willis Harman calls the "perennial wisdom."

I remain skeptical. By what new schema of human value are we to evaluate what it means when jacked-in twentysomethings gather in cyber-nightclubs only to talk to distant strangers in the lonely ether of cyberspace?

How should we interpret this fundamental shift in relationships between individuals? Is this the new masquerading as the strange, or vice versa? Is it an impulse toward modernism in a postmodern environment; an atavistic and misguided attempt to return to community; or the sign and symptom of a new and alienating society, in which all of our interactions will be mercilessly mediated by electronic filters?

In the mediated society, depersonalization is a powerful suasion. It is often (but not always) the result of technological forces. Face-to-face communication is devalued, but subtly rather than overtly. Attendant breakdowns in the civil order encourage these forms of communication as safer. Trust and dependency in human relationships are replaced by electronic security mechanisms of one sort or another.

In a mediated society, computer-mediated communication between strangers is held to be at least as valuable as or, in some cases, better than face-to-face communication within an established community. (That one could be blissfully unaware that one was communicating with an ax murderer online does not seem to bother some observers.) In the cyberculture mind-set, online transactions are perceived as purer, stripped of the messy complications of "meat space."

Assuming that we can make a case that we are steadily descending into both the mediated society and the new technocratic modes of control that might characterize it, we must take an honest look at what changes have already been wrought in our society and culture, especially over the last few decades. Many of these changes can be traced to the manifold impacts of the new utilitarianism. A useful barometer for the ground-level effects of these social changes is the family, traditionally considered the basic social unit of any community.

The dissolution of the family structure has been the subject of much ideological and political hand-wringing in recent years. However, much of the political dialogue misses the point and obscures the real dynamics involved. While conventional wisdom holds that the erosion of the family began in the sixties, in actuality it dates back to the beginning of the industrial revolution.

At that time, fathers were, in essence, removed from the family unit by irresistible economic forces as they headed off to work in William Blake's urbanized "dark Satanic mills." With the advent of public education in the nineteenth century, children were for the first time in history collectivized in schools. Education became a state function, and the modern educational system was born. One parent (the mother) became the sole anchor of domestic existence.

This dynamic has continued to work quietly since that time. With the advent of a superheated global economy, these social and economic forces have intensified their relentless onslaught against family and community. Kirkpatrick Sale's second industrial revolution and Toffler's Third Wave represent revisitations of history in which a confluence of economic forces has removed another parent—this time the mother—from the home and into the collectivized workplace.

The cycle of collectivization is complete when mother takes child into corporate day care. The family as a microcommunity becomes fully fragmented. In practical terms, the effects are easy to trace. Family members no longer share evening meals but eat on the run and then beat a hasty retreat to electronic stimulation for the rest of the evening.

I have painted a grim picture of life in the contemporary United States, but I think most readers will acknowledge that it is in many ways an accurate one. Television sets—no longer an electronic hearth for gathering together—have multiplied within the household. Just as there are personal computers, there are now personal televisions to reinforce the sad convenience of electronic segregation.

There are, of course, counterbalancing positives: for example, a flowering of personal expression and choice, which often treads a thin line between narcissism and fulfillment. The sight of a parent jogging by the side of the road with a toddler in a hand cart symbolizes a continuing quest for personal fulfillment that can be construed as a part of the true legacy of the sixties. There are other examples of the positive aspects of this trend. Nevertheless, we have yet to come to terms as a society with the long-term effects of this new evolving lifestyle, especially its effects on our children.

There are many who feel that the single most serious defect in the current social landscape is parents' inability to spend adequate amounts of quality time with their children. It has been well documented that the amount of time that adults spend with their children has steadily decreased over the last twenty years or so. It seems critically important to ask why and to probe for root causes.

As is often the case, the real answers are economic. Many aspects of this trend can be traced to the need to keep up with the demands of our ever accelerating society; this acceleration is driven by new communications efficiencies and the unyielding pace of life they engender. In any event, the disturbing disruption of parent-child relationships represents a fundamental breakdown in the cycle of human nurturance and bonding. These factors, along with high divorce rates and the implicit societal sanction of violence as a solution, unquestionably lie at the root of teenage violence and the rise of so-called superpredators who increasingly engage in what in the past were exclusively adult crimes.

Let us make no mistake, something is very wrong here; and our society, locked into in a dizzying postmodern tailspin with all the controls frozen, has yet to come to terms with the problem. In many cases, electronic stimulation—whether television, online services, or electronic gaming—is also sucked into the parental vacuum as a substitute for diminished intersubjectivity and interfamilial community. The *Wall Street Journal*, for example, has reported on the phenomenon of yuppie parents who take their children to shopping malls and then deposit them at Radio Shack, or a similar store where electronic games are available, so that they can shop unencumbered by their children, a practice that is a vivid example of Lasch's "culture of narcissism."

What is troubling here is that despite the obviousness of the problem's source, our society appears to be in serious denial concerning its root causes. The fact that these issues have moved into the political arena suggests that attempts to deal with them at the societal level have been largely unsuccessful. And yet, in some ways, the political arena is where the accountability for their origin belongs, since increasing corporate control over political mechanisms is now well documented.

Nevertheless, the random and ill-conceived political attempts to solve social problems that we are seeing today are bound to fail. In fact, these problems will only be aggravated by new forms of government intervention, especially with respect to children's issues. These include, for example, President Clinton's proposal to impose a national curfew for teenagers; the attempt on the part of elementary and secondary schools to teach values as opposed to basic skills, under the guise of educational reform; and the use of the V-chip to serve as a surrogate parent by screening television content for inappropriate sex and violence.

Of greatest concern is the "parenting break" described by both Robert Bly and William Irwin Thompson. Many child-care experts now believe that the day-care experiment has largely been a failure. Nevertheless, that new types of child-care centers and other questionable approaches to "surrogate parenting" are in the offing suggests that the current cycle of alienation and commodification of societal relationships is far from having run its course.

One of the latest trends in "upscale parenting," for example, is the rush to enroll toddlers in competitive early learning experiences where at the age of two or three they can begin to prepare themselves for some unspecified future role in the new corporatism. Given the current trajectory and the increasingly common commodification of reproductive "services," it seems only a matter of time before we begin to inch toward the disturbingly repressive scenarios outlined by B. F. Skinner in *Walden Two.*

The societal shift that has been described is, of course, a worst-case scenario based on a fairly linear projection of the future derived from currently observable social trends. But other options and possibilities may lie ahead. If R. D. Laing was correct in stating that societies can take wrong directions just as people do, then an important choice confronts us.

As technology pulls society in one direction, an "invisible" counterforce tugs quietly in another. Once again, we see the law of opposites at work. It is this trend—or perhaps megatrend—that represents a truly radical new vision of society in which technology must assume its proper place or else become devalued currency.

According to the late Willis Harman, a former Stanford Research Institute (SRI) researcher who headed up the Institute for Noetic Sciences in California, the creative wave that began in the sixties with a fundamental change in human perception of planetary priorities is alive and well. Its form, however, is not always easy to describe. Harman has pointed to the work of San Francisco–based sociologist Paul Rey, who statistically tracks the effects of long-term cultural transformation. According to Rey, there are now over forty million people in the United States who can be characterized as having a worldview that emphasizes ecological, spiritual, and matriarchal value systems.

Thus, the legacy of the sixties is still with us. It can be seen in the values of college and high school students. It can be seen as a growing trend in other sectors of society. It is this group and system of values—and not those of digital culture, with its convoluted posturings about Net spirituality—that can lay claim to this legacy.

It is the reaffirmation of this system of beliefs, bounded by a common and perennial wisdom, that will help us correct the current trajectory— the flight into the virtual—by grounding us in a more environment-oriented, ecological system of values. In reaffirming this system, we will have taken the first steps away from the alienating influence of a mediated society— a future we can avoid if we so choose—and toward a deeper sense of local and planetary community.

Orwell Reconsidered:

The Paradox

of Decentralization

Are we abusing the concentration of information that computers enable, or we are using them wisely? Should we be concerned about George Orwell's predictions that new forms of social control will arise coincident with these new technologies, or are such concerns merely overreaction? Are the dystopian social scenarios depicted in current science-fiction vehicles like *Blade Runner* or *Neuromancer* possible? Or will information technology, with its stunning powers of aggregation, create a world in which technology's capabilities and regimes are, in effect, used as weapons against the interests of ordinary citizens? Will a virtually cloaked oligarchy with Oz-like powers emerge, capable of manipulating the vast machinery of computer networks and the New Media for the purpose of ambitious and overarching social control?

Although these are far-fetched scenarios, they do raise issues that are more than mere idle concerns. Indeed, some of the trends depicted in these science-fiction visions are already happening to some extent. Media concentration, for example, is at an all-time high. During the last five years alone, primarily through mergers and acquisitions, the number of owners of large media establishments has dwindled considerably.

What effect will these new media empires have on the free flow of information in a democratic society? There are concerns that corporate interests are now far more well represented in the mainstream media than has traditionally been the case. Many media conglomerates are now owned by Fortune 500 companies with no traditional connection to the media business.

One trend that has emerged in this new landscape is a propensity toward what might be called "managed reality," perhaps a distant cousin of virtual reality. A good example is the Republican National Convention held in San Diego in August 1995. From a media standpoint, the convention was a watershed event. It was the first "virtual" political event of a scale and magnitude capable of making a lasting impression on the American public. As it was in the Gulf War, information presented to the American public was very tightly controlled and highly scripted. Events were specifically planned with an eye to how they would play on television. In other words, the convention was designed to be a media event, as opposed to an event that was covered by the media. The distinction is an important one. In a media event, viewers see only what media handlers want them to see. There is indeed a certain Orwellian quality to this kind of media manipulation, a fact that was not lost on many commentators, including NBC's Tom Brokaw, who complained bitterly about it in an op-ed piece in one of the major newsweeklies.

Media theorist Neil Postman has an interesting take on what might be called the Orwell question. In his book *Amusing Ourselves to Death,* Postman argues that it is not Orwell's vision of centralized domination and control that we need fear but rather Aldous Huxley's view as articulated in *Brave New World:* "What Orwell feared were those who would ban books. What Huxley feared was that there would be no reason to ban a book, for there would be no one who wanted to read one. Orwell feared those who would deprive us of information. Huxley feared those who would give us so much that we would be reduced to passivity and egoism. Orwell feared the truth would be concealed from us. Huxley feared the truth would be drowned in a sea of irrelevance."[3]

We can see the vague outlines of Postman's concern about amusing ourselves to death in emerging electronic bread-and-circuses scenarios of contemporary life, including MUDs, MUSEs, virtual reality, electronic gaming, and electronic gambling, which is one of the fastest growing businesses on the Internet. Furthermore, we can rightly wonder whether these diversions are the mark of a healthy society. Consider, for example, Katie Hafner's depiction of married life in the nin

> Wendy Metzler, a 23-year-old homemake
> computer-game widow. Her husband was

customer—for a company called the Total Entertainment Network (TEN) and had become addicted to a game called Duke Nukem 3D. For a month, he played it almost every night until morning. "I told him I was tired of sitting alone in bed," she says. Finally, two months ago, he persuaded Wendy to try logging on. Did she like it? "I haven't got off the thing yet," she whispers, sounding a bit embarrassed.

Now, once dinner is over and her 5-year-old daughter is in bed, Wendy sits down to her PC with enough Ho Ho's and chips to last her till dawn. Why the fascination? For Metzler, it's not just that Duke Nukem 3D, a gruesome shoot-'em-up where players navigate post-apocalyptic Los Angeles hunting down aliens, is fun. It's also that TEN links her with people through the Internet; that means Metzler (a.k.a. Daisy-Duke) can spend her nights meeting friends and new opponents. With screen names like Chen, Javamamma and HellKnight, her playmates challenge opponents who have logged in from places like Beaumont, Texas, and Short Hills, N.J. "Everyone knows me," she says.[4]

Adult preoccupation (and children's obsession) with electronic games suggests an odd kind of role reversal. Adults, under the influence of digital technology, revert to childish behavior, whereas children take on the worst and most aggressive traits of adults. Here again there are issues that revolve around electronic babysitting, the disruption of the nurturance cycle, and the abnegation of parental responsibility, wherein "The Child is father of the Man," but in a sense that is less-than-beneficial compared to Wordsworth's intention.

In this new scrambling of the "ages of man," traditional rites of adult passage are lost as the average age of puberty slides lower and lower. Such speculations bring to mind a chilling passage from Lewis Mumford's prophetic and intellectually adventurous *The Transformations of Man:*

> One cannot, of course, deny that large tracts of our lives have become increasingly bovine, vulpine, and simian. The age of the "men who are ten years old," long ago predicted in the Pali texts of Buddhism, is already visible. "There will be a time when children will be born to men who only reach an age of ten years old, and with these men, girls of five

will be fertile . . . with these men of an age of ten, violent hatred against each other will predominate, violent enmity, violent malevolence, violent lust for wholesale killing." [5]

If there are Orwellian dangers lurking in the shadows of the information revolution, how can they be reconciled with the benefits that computers and communications can yield? A fundamental dynamic in the cyberpunk mythos, the original fuel source for digital culture, was an information-age version of the David and Goliath story. Computer technology was the great leveler, and a rallying anthem of digital culture was the ability of information technology to empower and fortify the individual against the corporate megamachine. But the dystopian science-fiction scenarios favored by digital culture are by and large extreme depictions: the lone computer hacker flailing away at the system's entropic but formidable hegemony in a world gone wrong, essentially the black-and-white world of vivid but unfulfilled adolescent imagination. Unfortunately, digital culture has so far not come up with—and is not likely to—adequate societal models for the empowerment of individuals. Where are there opportunities for computers to empower truck drivers, grandmothers, schoolchildren?

For all their fascination with dystopian futurism, the thought leaders of digital culture remain somewhat naive about the real abuses that could occur in a fully "wired" society. Or, more precisely, their strongly libertarian values allow them to imagine abuses of power occurring within the domain of government while ignoring the fact that the potential for abuse by corporations is every bit as great.

It is curious how nonjudgmental—even passive—we have become about these concerns. Orwell, for example, is barely mentioned these days, and if he is, it is only in passing. In other venues, we are assured that such trends bear no relation to the grim visions of Huxley and Orwell.

Several years ago in a speech at a telecommunications conference called Supercomm, Peter Huber carefully explained why we are not becoming an Orwellian society; many of the arguments were likely drawn from his book *Orwell's Revenge*. Although not a member of digital culture per se,

Huber is squarely in the camp of Newt Gingrich, *Wired,* and George Gilder when it comes to communications policy, and he shares with them a certain obvious disdain for government involvement and proactive public policy. Huber's argument was a relatively simple one: Orwell's premise is based on the notion of centralized control of information, whereas the Internet and other digital technologies are distributed and decentralized.

While appealing on its surface, Huber's thesis has a basic flaw: Privacy can be abused in an environment characterized by a large number of distributed computers just as easily as it can be in a highly centralized model. In fact, in some ways the distributed model is even more problematic because privacy-violating information can exist throughout a networked grid and never be fully accounted for or precisely located. In addition, the case with which computers can be networked via the Internet means that information can be readily transferred, which renders the centralization issue moot.

In any event, corporations will indeed be tempted to abuse the new powers that these systems offer, as will governments and other entities. Privacy battles in these domains are being won and lost every day in the courts, as well as in courts of opinion. Some of the erosions are rather stunning—such as the creation of a national DNA data bank, already underway—but not as stunning as the public's numbness to the gradual diminishment of their civil rights.

Is Cyberspace

a Trojan Horse

for Technocracy?

Technocracy is not a new idea. The technocracy movement dates back to early-nineteenth-century philosopher Auguste Comte, the founder both of modern sociology and of positivism. The movement continued in the late nineteenth century with the work of Frederick Taylor, who developed principles of scientific management and techniques like time-and-motion studies for measuring and improving human output in the workplace. In the twentieth century, it again surfaced in various forms, including cybernetics, general systems theory, and even aspects of behaviorism that blossomed during the 1960s. There were attempts to formally establish technocracy during the 1930s, as documented by Jeremy Rifkin in *The End of Work*. Today, the notion that human behavior can be not only understood but also managed and improved upon via scientific methodology has been rekindled. For example, although chaos and complexity theory are promising scientific developments, they are already being coopted and adapted to the workplace to create more efficient business organizations.

The one common element in all these systems of thought is the belief that scientific principles can be applied to human activities in such a fashion as to better society, a belief that Neil Postman calls "scientism." In the current incarnation of technocracy, the proponents of digital culture are leading the charge with their belief that social and government mechanisms can be readily replaced by the Internet and that digital technology, in general, is an unstoppable force for social improvement.

But technocracy is not just about technology—it is also about the exercise of power. And a power shift has occurred among our major institutions—power has shifted not from church to government, as was the case in the last seismic shift, but rather from government to corporation. Since the computer and communications industries are among the most important sectors in the new Third Wave economy, it is not surprising that they are gaining significant influence over the machinery of government decision making.

What is troubling is that such activity seems to be taking place outside of the traditional process of government oversight, and thus there may be no effective forum available for debate and public deliberation. For example, Americans were not given a voice in the decision to distribute bioengineered foods in the marketplace. And it is unlikely that they will be given a chance to vote on a referendum concerning the use of widespread video surveillance in cities and towns. Nor will they be given a chance to debate whether the Internet should be used to teach their children in schools. Yet all of these changes are moving ahead with great momentum and will have profound effects on our collective future. Assuming the movement toward increased technocratic control continues, Americans may soon find they have very little say in some fundamental issues affecting their everyday lives.

If the American public does not make these decisions, then who will? There are an increasing number of public policy matters that concern the intersection of technology and societal issues. But because these matters have strong technological dimensions that are beyond the ken of the ordinary citizen, they are increasingly being implemented outside of the normal political process.

One of these public policy matters is education. The widespread concern about the deterioration of public education across the nation has opened the door for technocrats armed with the solution: hooking up to the Internet. (For the time being, let us put aside any specific judgment as to the efficacy of doing this—there are worthy arguments on both sides.) The point is that a power shift from traditional educators to technocrats has taken place, and the trend is likely to continue.

Increasingly, decisions affecting quality of life and lifestyle are being made not as a natural outgrowth of the political process but rather by the sheer momentum of the marketplace and the technological diaspora now underway. We should translate Nicholas Negroponte's statement that the computer revolution is "unstoppable" as saying that the pressures of technological innovation coupled with the gathering momentum of an unfettered marketplace will result in a managed (rather than governed) society in which the role of the Internet and companion technologies will be considered essential.

The Internet will become a Trojan horse for the advent of technocracy if in the process of embracing the new technology we are also tacitly required to embrace a new set of social constructs as well. One of the dominant messages of cyberculture is that the digital revolution is an entity, a gestalt, an all-of-a-piece phenomenon that must be rejected or accepted in toto.

This message is, of course, patently false. There is absolutely no reason why we must accept all types of technology if we accept one. Unfortunately, the choice is often framed as one between mindless Luddism and a bold, forward-thinking march into the information age and the twenty-first century, a false dichotomy that only confuses the issue.

In this, we can see the outlines of one of the more interesting paradoxes of the digital revolution: In order to understand the social implications involved, we must become part of the digital revolution, but once we do, we have already contributed to its forward progress. Hanging back in the shadows and accepting the hype at face value while not validating it through our own experience is not tenable. However, if we learn about the technology and use it, then we have crossed over the "virtual threshold."

I am convinced that becoming experientially informed and involved is the best option for two reasons. First, we cannot criticize what we do not understand; and second, we all need to learn how to be more discriminating about what has the potential to be a set of very useful technologies.

Science, Spirituality, and the Crisis of Epistemology

The struggle between secular scientism and spiritual renewal remains a dominant, but not highly visible theme in American culture. The outcome of this struggle is far from certain. As a method of empirical description, as a means of charting and discovering the elegant but often elusive hidden order of the natural world, and as an approach to bringing new efficiencies and technological advantages to human existence, Western science has an impressive list of achievements. But as a value system and a singular and exclusive epistemology, it seems to have run its course. As we cross the chaotic threshold of a new millennium, the old order gathers all of its strength for a coup de grâce that can never happen. From the elongated perspective of history, this is what might be called the supernova effect: Stars shine brightest just before they disappear.

We seem to be at a crossroads, a place where a new civilizational paradigm is emerging, although we are unsure what form it will take. Theologian Teilhard de Chardin saw it as a convergence of "science, religion, and philosophy." William Irwin Thompson and Lewis Mumford, both extraordinarily prescient cultural historians, foretold the development of a new world culture.

Unless the proponents of technocracy and digital culture are successful in selling their cleverly recycled scientific materialism, Western science, for all of its wonderful contributions, is likely to be replaced as a system of values. Interestingly, an odd and out-of-proportion manner of thinking about the computer and the Internet may be digital culture's attempt to stoke the dying embers of the old paradigm. The struggle between these and other forces is what defines the ambiguous and protean Sturm and Drang of the postmodern condition. And, in some ways, the fate of the planet, in millennial suspension, hangs in the balance.

There is commonplace hoping-against-hope mirage that a technological breakthrough will bail us out of these complications. However, this appealing but wayward notion may serve only to delay a necessary and still badly needed reordering of planetary priorities, a new phase in the evolution of mind and consciousness that began in the sixties but remains unfinished business.

Proponents of digital culture, such as Kevin Kelly and Howard Rheingold, who are pushing the notion that technology will provide a quick fix or magic cure for the earth's imperiled ecosystems have come up with a marketing formula that seems to work. But it works only by costuming the status quo in the garb of counterculture to peddle the shopworn messages of the old order, the dominant paradigm.

Perhaps the metaphysics of the Net theme that *Wired* supports was created with the knowledge that for secular scientism to succeed, it would somehow have to fill the void left by spirituality and other systems of value that offered genuine transcendence. But if so, an important principle has been forgotten: Technological powers and capabilities are only truly successful to the extent that they are fully humanized. When the process is reversed and our technologies begin to shape us in their image and likeness, we are heading in the wrong direction.

The Internet is an exciting new means of human communication. Its use will continue to enable important scientific breakthroughs (although some achievements, such as human cloning, may only create additional crises of values). In the promising but unstable environment that lies ahead, these breakthroughs may help us to deal with the reactions of the natural environment to our own metalevel meddling.

The Net may also help to transform old forms of social organization to pave the way for the reordered priorities of a new global order. But, contrary to the beliefs of digital culture, the Net itself is not what this coming transformation is all about; the Net is only a tool—albeit an important one—for dealing with the coming crisis of epistemology and ecology.

If we believe digital culture's message, we can become deflected and distracted from this reordering of priorities and the need to humanize technology through a new awareness. If this reordering does not occur, the new technocrats will ride the wave of the Net's upscale cachet in an attempt to impose their reductionist views on the protean nature of the post-modern condition, caught as it is in the nebulous transition between old and new.

If technocracy is camouflaged as hip new technologies, then there will likely be more-subterranean conflict as the technological imperative continues its secularizing influence. The cultlike quality of digital culture will continue to absorb and coopt the more genuine spiritual and human-istic impulses that arise in reaction to it, creating a false rapprochement. Under the banner of *Wired*'s vaguely defined Net metaphysics, digital cul-ture as the upscale vanguard of the new technocracy may attempt to offer more-absurd amalgamations of science and spirituality in the attempt to provide a sense of the transcendent where none can exist.

For nouveau scientific materialism to succeed as a replacement for spirituality, it will have to adopt a mock-spiritual fervor of expression to further disguise the confusing inversions of spiritual and material values. If the intention is ultimately for technology to replace religion and phi-losophy, then it must purport to fulfill the human need for transcendence that the latter have traditionally addressed. It might seem that only the culturally desperate would latch on to such an ontological scam, but there is a kind of cultural desperation in the air these days.

There are those who argue that the opposition between science and spirituality has, if anything, diminished over the years. This is an argument worthy of consideration. In the pure world of academic discourse, a num-ber of trends seem to suggest this. Proponents point to such phenomena as the Gaia hypothesis in biology; the new physics; fuzzy logic; and the increasing influence of chaos and complexity theory on scientific formu-lations—all basically attempts to accommodate and describe nonlinear systems behavior (which, of course, includes human behavior) in a more generalized set of principles.

In a way, such attempts are not surprising. Science seems to sense its difficult and precarious ontological state in the postmodern world and has scrambled to retheorize and widen its metaphysical borders. But while in a purely abstract and theoretical sense, it could be conceded that there are indeed scientific and nonscientific intellectual traditions that can justifiably reconcile their long-standing mutual exclusivity in the attempt to find common ground, there are other factors that militate against too facile an acceptance of this "new science."

For one thing, the noble, traditional use of science as a means of pure exploration has been eroding for some time. This erosion complicates the picture immeasurably. Under the influence of the new utilitarianism, academic science is no longer content to explore purely theoretical issues but is increasingly being asked to shift its research into the more immediate mode of the practical and pragmatic.

While this shift is not all bad, it has downsides that almost do not need enumeration because the media serve them up regularly: the falsification of research results because of commercial pressures; the politicization of science results by government and corporate interests; the ever increasing dependence of universities on corporate largesse; and the growing pressure on science in general to make its activities profitable in shorter and shorter cycles of research. Theoretical constructs are nice, but in the real world science now is marching to a different drummer: the interest of business and commerce.

In the accelerated marketplace shaped by the new utilitarianism, not only has technology been put to work to develop the ever vaunted competitive edge but also the omnivorous hunger for new solutions has foreshortened and diverted the cycle of research and development. The distinction between pure and applied science has been blurred, and much of the work of theoretical science is being robbed in the cradle before there is adequate time to see it to fruition in the nobler traditions of pure scientific research.

To some extent, science and technology are diminishing their own credibility by allowing these misguided applications toward social ends. In the computer age, cybernetics and general systems theory have been reborn and are redirecting their focus toward social concerns for which their proponents feel they will have practical application. Kevin Kelly, in fact, heralds the rebirth of cybernetics in *Out of Control.*

Overlaying these new approaches is a confused metaphysical cooptation that Morris Berman calls "cybernetic holism." In his book *Coming to Our Senses,* Berman warns about its dangers: "Hence if we are to look for possibilities for co-optation, we must think of scientific/industrial/corporate co-optation. . . . the somatic energy of holistic thinking becomes the conceptual structure of cybernetics, or systems theory . . . that, not a new Christianity or fascism, is the real threat facing us today."[6] I believe that Berman is warning us about a new modality that results from an odd hybridization of science and spirituality that distorts the true metaphysical nature of both.

If science and spirituality are truly to be reconciled, then in the realm of possibility there is both an end result which is genuine as well as anomalous accommodations that could occur. Once powerful technological memes are harnessed to the large engines of politics and commerce, they can be easily coopted, inappropriately morphed, and roundly misinterpreted.

Berman's cybernetic holism seems to be an example of an inappropriate fusion between otherwise worthwhile concepts, an unholy alliance rather than a transformational reintegration. I would suggest that the concept of cybernetic holism could easily apply to the amateur metaphysics that is the stock-in-trade of *Wired,* a metaphysics to which our society, in its current modality, seems a bit too susceptible.

The evolutionary pretensions of digital culture point toward some vague sort of apotheosis of computer intelligence. But the proponents of digital culture want to do this without the encumbrances of the past: literature, philosophy, spirituality, and other traditions. They want to strip away the past to create a new vision of humankind, loosely based on the technological imperative. However, without the humanistic markers and guideposts afforded by the traditions of the past, it seems all too easy to wind up in a curious no-man's-land of cultural and intellectual relativism—not the positive and creative state of ambiguity afforded by postmodern thought but an elasticity of values that is more characteristic of nihilism.

In this environment, figure and ground are indistinguishable, and meaning has been badly eroded: Spirituality can don the guise of materialism, or vice versa, and there is no way to tell the difference. The result is a strange hybrid of spirituality and materialism that entices and confuses: dehumanization with a happy face.

From the standpoint of digital culture, the phenomenon seems to have more to do with sacralizing science than with secularizing religion. From the standpoint of technocracy in the larger sense, the opposite is true. But if science and technology are going to provide the ultimate answers to fundamental human questions, they will have to work much harder. If all the mythos of cyberculture can do is mimic the religious quest for "eternal life" in the absurd technological guise of a bad science-fiction plot—the downloading of consciousness into a presumably immortal computer, which is to say the achievement of the objectives of religion by nonreligious means—then that is a poor offering indeed.

As we approach the uncharted terrain of the new millennium, instabilities on the surface of the planet mirror the uncertainties and complexities of our postmodern condition. Our very own metaphysical ground of being is being altered, and reality itself is up for grabs. Interestingly, the notion of virtual reality is appearing at a time when epistemological uncertainties are rampant. In one way of looking at it, virtual reality is at war with consensus reality, and the former has the temporal advantage.

As we ponder these complexities, it is worth wondering whether virtual reality is more useful as a metaphor than as a technology. We might also wonder whether the true gift of computers and communications might not be an epistemological one: They have allowed us to question and deconstruct the premises of our sense of reality.

Only in this limited sense does the culture they have spawned share some commonalities with the deeper existential questioning characteristic of the sixties. One of the principal discoveries (or rediscoveries) made in that unique period in human history was this: We create our own realities.

In the new environment, we will, unfortunately, no longer have the luxury of using science as the ultimate arbiter of reality. The new landscape will be far more complex and nuanced. But neither formulaic religion nor spirituality can alone fill the void. What seems necessary for this transformation is a new model for transcendence, a model that takes into account the more arbitrary dimensions of the human condition. In this sense, postmodernism must be seen more as a starting point than as an end result.

The conflict that we have witnessed throughout the turbulent nineties is largely related to a breakdown not just in the civil order of society—after all, this is only a symptom—but more importantly in the very structure of knowledge itself. The process is being rendered even more complicated as knowledge itself becomes paradoxically commoditized.

We are experiencing an epistemological crisis, and many of the new digital technologies are at the very center of it. A false rapprochement between science and spirituality will only exacerbate this crisis. Looking for a way to truly reconcile these very different ways of knowing will be the challenge of the new millennium.

Notes

----- Virtual Dreams

1. Tom Valovic, "A Giant Step towards Internet Commercialization?" *Telecommunications*, June 1991.
2. Sven Birkerts, *The Gutenberg Elegies: The Fate of Reading in an Electronic Age* (Boston: Faber and Faber, 1994), 216.
3. Alvin Toffler, *Powershift: Knowledge, Wealth, and Violence at the Edge of the Twenty-First Century* (New York: Bantam Books, 1990).
4. Quoted in review of James Burke and Robert Ornstein, *The Axmaker's Gift: A Doubled-Edged History of Human Culture, Publisher's Weekly*, August 7, 1995, 45.
5. Gary Chapman, "Barbed Wire: Is the New Cyber-Mag Any Good?" *The New Republic*, January 9, 1995.
6. Kevin Kelly, *Out of Control: The Rise of Neo-biological Civilization* (Reading, Mass.: Addison-Wesley, 1994), chap. 2.
7. "The Promise and Peril of Emerging Technologies: A Report on the Second Annual Roundtable on Information Technology," August 4–8, 1993, The Aspen Institute, p. 4.

----- Virtual Nightmares

1. John Brockman, "Agent of the Third Culture: John Brockman Is the Michael Ovitz of the New Intellectual Elite," *Wired*, August 1995, 44.
2. Joshua Cooper Ramo, "Finding God on the Internet," *Time*, December 16, 1996.

----- The Electronic Polity

1. Thomas Valovic, "Online Encounters," *Media Studies Journal* (spring 1995).
2. Charles Swett, *Strategic Assessment: The Internet* (paper prepared for the Department of Defense, July 17, 1995, and posted on the Internet by the Project on Government Secrecy of the Federation of American Scientists, http://www.fas.org/cp/swett.html).
3. Howard Rheingold, *The Virtual Community: Homesteading on the Electronic Frontier* (Reading, Mass.: Addison-Wesley, 1993), 14.

----- Digital Culture

1. Paulina Borsook, "Cyber-Selfish," *Mother Jones,* July–August 1996, 59.
2. Ibid.
3. Stewart Brand, *The Media Lab: Inventing the Future at MIT* (New York: Viking, 1987), 7, 9.
4. Kevin Kelly, *Out of Control: The Rise of Neo-biological Civilization* (Reading, Mass: Addison-Wesley, 1994), 257.
5. Stewart Brand, *Whole Earth Catalog,* 1st ed., 1972, pp. 1–2.
6. Joshua Quittner, "The Merry Pranksters Go to Washington," *Wired,* June 1994.
7. Mark Dery, *Escape Velocity: Cyberculture at the End of the Century* (New York: Grove Atlantic, 1997), 93.
8. Charles Swett, *Strategic Assessment: The Internet* (paper prepared for the Department of Defense, July 17, 1995, and posted on the Internet by the Project on Government Secrecy of the Federation of American Scientists, http://www.fas.org/cp/swett.html).
9. Borsook, "Cyber-Selfish," 59.
10. Nancy Gibbs and Karen Tumulty, "Master of the House," *Time,* August 12, 1996.

----- Science, Culture, and the Internet

1. Christopher Lasch, *The Culture of Narcissism: American Life in an Age of Diminishing Expectations* (New York: Norton, 1979), 20.
2. Christopher Lasch, *True and Only Heaven: Progress and Its Critics* (New York: Norton, 1991).
3. Neil Postman, *Amusing Ourselves to Death* (New York: Penguin, 1985), vii.
4. Katie Hafner, "Log On and Shoot," *Newsweek,* August 12, 1996. © 1996 Newsweek, Inc. All rights reserved. Reprinted by permission.
5. Lewis Mumford, *The Transformations of Man* (New York: Norton, 1991), 133.
6. Morris Berman, *Coming to Our Senses: Body and Spirit in the Hidden History of the West* (New York: Simon and Schuster, 1989).

Bibliography

Ben-Tov, Sharona. *The Artificial Paradise: Science Fiction and American Reality.* Ann Arbor, Mich.: University of Michigan Press, 1995.

Berman, Morris. *Coming to Our Senses: Body and Spirit in the Hidden History of the West.* New York: Simon and Schuster, 1989.

Birkerts, Sven. *The Gutenberg Elegies: The Fate of Reading in an Electronic Age.* Boston, Mass.: Faber and Faber, 1994.

Bly, Robert. *The Sibling Society.* Reading, Mass.: Addison-Wesley, 1996.

Borsook, Paulina. "Cyber-Selfish." *Mother Jones,* July–August 1996.

Brand, Stewart. *The Media Lab: Inventing the Future at MIT.* New York: Viking, 1983.

Brockman, John. "Agent of the Third Culture: John Brockman Is the Michael Ovitz of the New Intellectual Elite." *Wired,* August 1995.

Burstein, Daniel, and David Kline. *Road Warriors: Dreams and Nightmares along the Information Highway.* New York: Dutton, 1995.

Chapman, Gary. "Barbed Wire: Is the New Cyber-Mag Any Good?" *The New Republic,* January 9, 1995.

Dery, Mark. *Escape Velocity: Cyberculture at the End of the Century.* New York: Grove Atlantic, 1997.

Dyson, Esther. *Release 2.0: A Design for Living in the Digital Age.* New York: Broadway, 1997.

Hafner, Katie. "Log On and Shoot." *Newsweek,* August 12, 1996.

Huber, Peter. *Orwell's Revenge.* New York: Free Press, 1994.

Kellner, Douglas. *Jean Baudrillard.* Stanford, Calif.: Stanford University Press, 1989.

Kelly, Kevin. *Out of Control: The Rise of Neo-biological Civilization.* Reading, Mass.: Addison-Wesley, 1994.

Lasch, Christopher. *The Culture of Narcissism: American Life in an Age of Diminishing Expectations.* New York: Norton, 1979.

———. *True and Only Heaven: Progress and Its Critics.* New York: Norton, 1991.

Mandel, Thomas, and Gerard Van der Leun. *Rules of the Net: Operating Instructions for Human Beings.* New York: Hyperion, 1996.

Mander, Jerry. *In the Absence of the Sacred.* San Francisco, Calif.: Sierra Club Books, 1991.

Mumford, Lewis. *The Transformations of Man*. New York: Norton, 1991.

Negroponte, Nicholas. *Being Digital*. New York: Alfred A. Knopf, 1995.

Postman, Neil. *Amusing Ourselves to Death*. New York: Penguin, 1985.

———. *Technopoly: The Surrender of Culture to Technology*. New York: Alfred A. Knopf, 1992.

Quittner, Joshua. "The Merry Pranksters Go to Washington." *Wired*, June 1994.

Ramo, Joshua Cooper. "Finding God on the Internet." *Time*, December 16, 1996.

Rheingold, Howard. *The Virtual Community: Homesteading on the Electronic Frontier*. Reading, Mass.: Addison-Wesley, 1993.

Ronell, Avital. *The Telephone Book: Technology, Schizophrenia, and Electric Speech*. Lincoln: University of Nebraska, 1989.

Rushkoff, Douglas. *Media Virus: Hidden Agendas in Popular Culture*. New York: Ballantine Books, 1996.

Stoll, Clifford. *Silicon Snake Oil: Second Thoughts on the Information Highway*. New York: Doubleday, 1995.

Swett, Charles. *Strategic Assessment: The Internet*. Paper prepared for the Department of Defense, July 17, 1995, and posted on the Internet by the Project on Government Secrecy of the Federation of American Scientists (http://www.fas.org/cp/swett.html).

Tapscott, Don. *The Digital Economy: Promise and Peril in the Age of Networked Intelligence*. New York: McGraw-Hill, 1996.

Thompson, William Irwin. *The American Replacement of Nature: The Everyday Acts and Outrageous Evolution of Economic Life*. New York: Doubleday, 1991.

———. *The Time Falling Bodies Take to Light: Mythology, Sexuality, and the Origins of Culture*. New York: St. Martin's Press, 1981.

Toffler, Alvin. *Powershift: Knowledge, Wealth, and Violence at the Edge of the Twenty-First Century*. New York: Bantam Books, 1990.

Turkle, Sherry. *Life on the Screen: Identity in the Age of the Internet*. New York: Simon & Schuster, 1995.

Valovic, Thomas. "A Giant Step towards Internet Commercialization?" *Telecommunications*, June 1991.

———. *Corporate Networks: The Strategic Use of Telecommunications*. Norwood, Mass.: Artech House, 1992.

———. "Online Encounters." *Media Studies Journal* (spring 1995).

Index

About the Author

Thomas S. Valovic is a media theorist and a recognized expert in telecommunications. He has written extensively on the business and social impacts of computers and communications for a variety of publications including *Computerworld, PC Week, Information Week, Whole Earth Review, Media Studies Journal*, the *San Francisco Examiner*, and the *Boston Globe*. He is also the author of *Corporate Networks: The Strategic Use of Telecommunications*. Valovic is currently a research manager with International Data Corporation and a member of the adjunct faculty at Northeastern University. Prior to that he was editor in chief of *Telecommunications* magazine.